1000 Best

MOVIES ON VIDEO

CLASSICS FOR TV

D1434082

Hamlyn

ISBN 0 600 50013 6
1000 Best Movies on Video
was devised and produced by
Siron Publishing Ltd
20 Queen Anne Street
London W1N 9FB
Editor: Nicholas Keith
Production: Cindy Wallace
Designer: Ann Doolan
Researcher: Pepsy Dening

Published 1984 by
Hamlyn Publishing,
a division of the Hamlyn Publishing Group Ltd,
Astronaut House, Hounslow Road,
Feltham, Middlesex

Typeset by V & M Graphics Ltd, Aylesbury, Bucks
and printed by
Printer Industria Gráfica SA
Cuatro Caminos, Apartado 8, Sant Vincenç dels Horts,
Barcelona, Spain DLB 29969-1984

Acknowledgements

The National Film Archive, a division of the
British Film Institute, was the largest single
source of photographic material.
   Many of the illustrations come from stills
and other material issued to publicise
videos distributed by those companies
listed on pages 284–287. We apologise in
advance for any unintentional omission
and will be pleased to insert the appropri-
ate acknowledgement in any subsequent
editions.

We are particularly grateful to the following
for their help:

Guild Home Video
Longman Video
Videon (220 Wandsworth Bridge Road,
   London SW6)
Videospace Ltd
The Video Palace (100 Oxford Street,
   London W1)

Contents

Introduction

One prediction missing from George Orwell's apocalyptic vision of 1984 was that by the end of that year four British homes in ten would have a video cassette recorder (VCR) with a choice of more than 6,000 feature films for viewing in the home. In fact, the video has not reached anything like saturation and one day a video player may be as common as a television set; indeed the two will probably be offered as a single unit.

Whether this will mean the end of the cinema in its traditional form is a lively topic for debate. Already more is being spent on video each year, both hardware (the machines) and software (the tapes), than on going to the pictures; and if a family can watch a feature film more cheaply and more comfortably at home, with no meal to buy and no parking space to find, it is hard to see how the cinema as a bricks and mortar institution can long survive on its present scale.

The case for the continuation of the cinema is nonetheless a strong one because it is theatrical exhibition that gives a film its launch pad. Most people first hear about a new film when it is introduced in the cinema, with the attendant publicity and media coverage. The video distributor knows very well that if the cinema publicity is effective, his job is half done. What used to be 'the film of the book' is now 'the video of the film'. However it is likely that, as the video market develops and the thirst for new material increases, more films will be made specifically for home viewing, by-passing expensive cinema release, not only on video but through cable and satellite television.

Already the choice of films available on video is a bewildering one and it is not made easier by the fact that video dealers – for understandable commercial reasons – tend to limit their stock to a relatively small number of potentially popular titles and ignore the rest. On the other hand, if all 6,000 films were suddenly to be assembled under one roof, the customer would very soon realise how worthless so many of them are. If anyone did decide to watch 6,000 films, say at the rate of one a day, it would take more than 16 years.

The purpose of this book is to signpost a way through the maze by offering a critical guide to the best. The selection of 1,000 titles is necessarily a subjective one. There is room for argument both about those films included and those left out; everyone has their own favourites. What the book

PARAMOUNT PICTURES Presents A LUCASFILM LTD Production
A STEVEN SPIELBERG Film
Starring HARRISON FORD
KAREN ALLEN · PAUL FREEMAN · RONALD LACEY · JOHN RHYS-DAVIES
DENHOLM ELLIOTT · Music by JOHN WILLIAMS
Executive Producers GEORGE LUCAS and HOWARD KAZANJIAN
Screenplay by LAWRENCE KASDAN · Story by GEORGE LUCAS and
PHILIP KAUFMAN · Produced by FRANK MARSHALL
Directed by STEVEN SPIELBERG

COMING to a theatre near you in Summer 1984

INDIANA JONES
and the
TEMPLE OF DOOM ™

can claim is to have rounded up all the outstanding films on video and to offer a generous pick of the rest.

Ten categories

The titles have been divided into ten sections:
- Action/adventure
- Adult
- Children
- Comedy
- Drama
- Horror
- Musicals
- Science fiction/fantasy
- Thrillers
- Westerns

There are two reasons for doing this, rather than simply listing the 1,000 films from A to Z: firstly it enables devotees of particular genres, such as Horror films, Westerns or Musicals, to find more easily what is available in those fields. Secondly, the categories correspond fairly closely to those used in video catalogues and shops.

Any such division must in some cases be arbitrary. The Marx Brothers obviously belong under 'Comedy' but some readers may be surprised to find Polanski's *Cul-de-Sac* alongside *Duck Soup*. The justification is that comedy takes many forms and the black humour of *Cul-de-Sac* is one of them. But for those who might find the categories puzzling, here are some guiding principles.

Action/adventure. This section includes war films, swashbucklers, historical epics, disaster movies, martial arts; drama on a broad canvas.

Adult. The term is usually employed in the video world to denote 'sex' films of a more or less exploitative kind. The use of the label here is somewhat different, being an attempt to list films whose treatment of sexual matters may be sensational or explicit but which can also claim some artistic merit.

Children. Films made primarily with children in mind, although adults can enjoy many of them as well. Other films suitable for children may be found elsewhere, especially under Comedy and Musicals. A separate list of **Best Family Films** is given on page 280.

Comedy. A wide range is covered – from slapstick silents to black comedy and satire. This section also includes parodies, so look for *Blazing Saddles* and *Cat Ballou* here, not under Westerns.

Drama. This is by far the biggest category, but it may be useful to think of Drama as the obverse side to Action/ adventure, in other words human and domestic stories rather than global ones.

Horror. The usual array of monsters and witch-craft and other manifestations of the supernatural. Films like Hitchcock's *Psycho* and *The Birds* have been included here rather than under Thrillers because their effect is horror-inducing rather than merely to create suspense.

Musicals. Films in which the songs and/or dance routines are an integral part of the telling of the story.

Science fiction/fantasy. Stories about an unreal world, as opposed to the real one, and particularly the imagined world of the future, e.g. *Superman* and *Star Wars*.

Thrillers. Gangster films, police procedurals, domestic crime and spies from Richard Hannay to Harry Palmer and James Bond.

Westerns. Movies inspired by the settlement of the American West during the nineteenth century, including the European-made 'spaghetti' type.

If you still cannot find what you are after, there is an index to all the titles at the end of the book.

Series and sequels

One problem was how to deal with series of films (like the Carry Ons, the Doctors and the Pink Panthers) and sequels (*Jaws II*, *Rocky II* and *III*, *Superman II* and *III* and so on). For the Pink Panther and Doctor films it was decided to give a full entry to the first one – as being the best – and titles only of the others. For the Carry Ons there is a different approach: an entry about the genre as a whole, together with a list of the available titles.

The treatment of sequels was decided on individual merit. Films like *The Godfather: Part II* and *The French Connection 2* have value in their own right and rate a separate entry while an inferior spin-off such as *Jaws II* gets a brief mention under the entry for the original.

Classics

The reader will notice that 20 of the 1,000 films have been allocated more space. These have been picked out as outstanding of their kind, true cinema classics. Again the choice must be subjective and should not be taken as a last word: 20 other titles could have equal claim to classic status. Something for friendly argument. If there are objections that all the chosen 20 were made before 1970, the answer must be that classics take time to mature. A list of these will be found on p. 282.

Star ratings

Each entry combines several items of factual information, as well as a critical guide. The following star system has been used to highlight the exceptional films:

    *** Not to be missed
    ** Highly recommended
    * Well worth watching

The brief description in each entry is also intended to convey something of the film's quality, as well as giving an idea of its plot or theme. (The absence of a star does not mean that the film is worthless.)

Country and Year

The title of each film is followed by: country of origin and year of production.

Running time

The *video* running time is slightly shorter than the cinema running time because television shows films at 25 frames per second and the cinema at 24 frames per second. The video version of a film is usually unedited.

Symbols of Suitability

While it is hoped that none of the 1,000 films in this selection can be termed a "video nasty", there are titles that are not suitable for showing to children. The Video Recordings Bill, which became law in July 1984, makes it mandatory for all video cassettes to be submitted to the British Board of Film Censors (BBFC) for certification; and only those programmes passed by the censor will be legally available through video shops. As this certification process will take some time, we have devised our own system of guidance, to give a general indication of a film's suitability in terms of the appreciation and understanding of younger viewers.

Key

    □   suitable for *all* the family
    v   vetting by parents advisable
    ●   adults only (mature teenagers may enjoy some of these films but parents are advised to vet them with particular care)

Of course, it is hard to generalise and it is up to parents to decide. To err on the safe side, some films which are widely watched and enjoyed by all ages (and indeed we have them included in our 'Best Family Films' list on pp 280) have been given a 'v' symbol – e.g. *Raiders of the Lost Ark* and the James Bond movies. Although young children will probably enjoy them in parts, they may be bored by the

Includes the
hit single
"Flashdance – What
a feeling" by Irene Cara

complexities of the plot and there is also the question of violence.

In each case, the main actors are listed and then the director. At the back of the book there is an index to stars and one to directors so that anyone with a special interest in John Wayne Westerns or Hitchcock Thrillers can find a convenient reference.

Distribution

The name of the video distributor is the first entry on the bottom line. This should help anyone to track down a cassette which is not available in the local video shop. Those who have difficulty in obtaining any of the films listed are recommended to contact the distributor (full names and addresses are given on pp 284) who should be able to supply details of the nearest stockist and may make special arrangements to get the tape to you.

Part of the value of this book, it is hoped, will be in drawing attention to films of merit that might not be easily found in the average high street video outlet. Already in the short history of video in this country, titles have been withdrawn from catalogues because the distributor has considered them lacking in commercial appeal. Had their availability been more widely known, they might still be there. The answer to the retailer's response "there's no demand" is to create that demand. On page 282 you will find under "Pot luck" a list of interesting titles, which are no longer officially available but which may still be found in a few video shops. (Note: Titles are withdrawn for several reasons not just lack of commercial success).

Video Systems

There are three tape systems and they are not interchangeable. In other words, you cannot play a cassette designed for one system through the VCR designed for another.

The three systems are:
VHS – pioneered by the Japanese company JVC
Betamax ('Beta') – pioneered by the Japanese company Sony
Video 2000 ('V2000') – pioneered by the Dutch company Philips

In Britain, partly because it is favoured by the big television and video rental chains, VHS is the dominant format with around 73% of the market. Betamax has something like 23 per cent and Video 2000 the remaining 4 per cent. Almost all feature film cassettes come in VHS and Betamax versions but only one title in six is on V2000.

There are also two video systems which use not cassettes but discs:

LaserVision – developed by Philips

CED Videodisc – a collaboration between Hitachi of Japan, which makes the disc players, and the American company RCA, which produces the software.

So far the disc systems, being very new, have had only a small impact in Britain and the number of titles available for them is limited. However, the quality of recording is superior to tape and the prices of the discs much lower.

Retail price

It is true that 95 per cent of videos are rented, not bought, and there are clear reasons for this. One is that many people are content to see a film through once, or possibly twice, and then change it for another one, like a library book. That is always likely to be the case. The second point is that most cassettes have been far too expensive to buy, with prices up to £50 or even higher.

More recently, however, some video companies have been making a determined effort to encourage purchase, both by reducing the prices of existing titles and launching new ones at special low prices. There is now a good selection of titles – among them the sort of classic film that is worth watching over and over again – available at around £20. It seemed useful, therefore, to offer a simple price code

A   £30 or more
B   £20–£30
C   Below £20
R   Rental only

One word of warning is that there are no fixed retail prices for videos and there may be variations from dealer to dealer. The prices apply to VHS and Beta. V2000 tapes tend to be in the £20–£30 range, while discs are much cheaper – all titles at around £11 or £12.

With new video titles appearing month by month, and given the inevitable gap between copy deadline and publication, a book like this cannot pretend to be completely up to date. What it can claim is that, by picking out the best, it contains a high proportion of titles that would belong permanently in any top 1,000 list, irrespective of what is released in the near, and even not so near future.

Now to the important part of the book – and happy viewing!

Peter Waymark
August 1984

**KIRK DOUGLAS · LAURENCE OLIVIER
JEAN SIMMONS · CHARLES LAUGHTON
PETER USTINOV · JOHN GAVIN
and TONY CURTIS as Antoninus**

Directed by STANLEY KUBRICK · Music Composed and Conducted by ALEX NORTH
Screenplay by DALTON TRUMBO · Based on the novel by HOWARD FAST
Produced by EDWARD LEWIS · Executive Producer KIRK DOUGLAS
A Bryna Production . A UNIVERSAL RELEASE

Above Us the Waves
GB   1955   96m   v   bw

*John Mills, John Gregson, Donald
Sinden, James Robertson Justice
D, Ralph Thomas*

Allied operation to destroy the German
battleship *Tirpitz* in a Norwegian fjord with
midget submarines. With its stiff upper lip
heroics and general air of decency, a typical
British war film of its time. Good battle
sequences.
Rank   VHS, Beta   A

Aces High
GB   1976   110m   v   colour

*Malcolm McDowell, Christopher
Plummer, Simon Ward, Peter Firth
D, Jack Gold*

First World War drama which transfers
the plot of *Journey's End* from the trenches
to the air. McDowell as the new recruit to a
squadron in France whose commanding
officer – his schoolboy idol – is cracking up
under the strain.
Thorn EMI   VHS, Beta   A

The Adventures of Robin Hood***
US   1938   102m   □   colour

*Errol Flynn, Olivia de Havilland, Basil
Rathbone, Claude Rains
D, William Keighley and Michael Curtiz*

Hollywood spectacle at its best; a fast-
moving collage of galloping horses, clashing
swords and zinging arrows, filmed in rich
Technicolor and with a stirring Erich
Korngold score.
Warner   VHS, Beta, CED   A

All Quiet on the Western Front**
US   1930   103m   v   bw

*Lew Ayres, Louis Wolheim, Slim
Summerville, John Wray
D, Lewis Milestone*

Potent pacificist tract about young Germans going off to fight for their country in
the First World War and having their
idealism shattered by the carnage. Fine
achievement of early sound cinema.
CIC   VHS, Beta   B

Apocalypse Now*
US   1979   148m   ●   colour

*Marlon Brando, Robert Duvall, Martin
Sheen, Frederic Forrest
D, Francis Ford Coppola*

Sheen as a young officer in the Vietnam
war whose mission is to "terminate" a
rogue colonel (Brando) fighting his own
battles in Cambodia; ambitious, often
powerful, sometimes confused statement
about Vietnam.
CIC   VHS, Beta   A

13

Attila the Hun
Italy/France 1954 88m v col

Sophia Loren, Anthony Quinn, Irene Papas, Henri Vidal
D, Pietro Francisci

Quinn hamming it up as the leader of the barbarous Hun tribes against the crumbling Roman Empire and being tempted by the young Sophia Loren. Routine Continental epic with strong battle scenes and unintentionally funny dialogue.
Videomedia VHS, Beta, V2000 A

Back to Bataan
US 1945 91m v bw

John Wayne, Anthony Quinn, Beulah Bondi
D, Edward Dmytryk

US colonel (Wayne) leads a guerilla army in the Philippines after the fall of Bataan to the Japanese; efficiently made with strong battle scenes.
Kingston VHS, Beta B
with *Roadblock*, 1951 thriller (70m)

Barabbas
Italy 1962 144m v colour

Anthony Quinn, Arthur Kennedy, Ernest Borgnine, Valentina Cortese
D, Richard Fleischer

Written by Christopher Fry and Nigel Balchin, it boasts a heavyweight cast and some impressive spectacle. The plot concerns a thief pardoned instead of Jesus who goes to the silver mines and becomes a gladiator.
RCA/Columbia VHS, Beta A

Battle of Algiers**
Algeria/Italy 1965 135m ● bw

Jean Martin, Yacef Saadi, Brahim Hoggiag, Tommaso Neri
D, Gillo Pontecorvo

A sharply observed, almost documentary study of the bitter war which ended in Algeria's independence from French rule in 1962; banned for some years by the French Government but much more than just anti-colonial propaganda.
Capstan VHS, Beta, V2000 A

The Battle of Britain
GB   1969   110m   □   colour
*Laurence Olivier, Robert Shaw, Michael
Caine, Christopher Plummer
D, Guy Hamilton*

The reconstruction of the stirring events of
1940 should have been more engrossing,
given its heavyweight cast (though there is
a fine portrait by Olivier of Sir Hugh
Dowding); spectacular, if repetitive, aerial
sequences.
Warner   VHS, Beta   A

The Battle of the Bulge*
US   1965   135m   v   colour
*Henry Fonda, Robert Shaw, Robert Ryan,
Telly Savalas, Dana Andrews
D, Ken Annakin*

The Ardennes campaign of December
1944, with Fonda leading for the Allies and
Shaw as the crack German panzer commander; well-handled war spectacle in
which the real stars are the tanks and the
characters take second place to the action.
Warner   VHS, Beta   A

The Battle of the River Plate
GB   1956   114m   □   colour
*John Gregson, Anthony Quayle, Peter
Finch, Bernard Lee
D, Michael Powell and Emeric Pressburger*

How the German battleship *Graf Spee* was
scuttled in Montevideo harbour in 1939.
Finch's sympathetic performance as the
Graf Spee's captain is a highlight of the
film, which sticks faithfully to the real
events but lacks a coherent narrative.
Rank   VHS, Beta   A

Ben Hur*
US   1959   209m   v   colour
*Charlton Heston, Haya Harareet, Jack
Hawkins, Stephen Boyd, Hugh Griffith
D, William Wyler*

Rivalry between a Jew, Ben Hur, and a
Roman tribune, Messala, at the time of
Christ comes to a climax in a brilliantly
mounted chariot race. Then the most
expensive film ever made and winner of a
record 11 Oscars.
MGM/UA   VHS, Beta V2000   B

The Big Boss
Hong Kong   1971   95m   ●   colour
*Bruce Lee, Maria Yi Yi, James Tien
D, Lo Wei*

The film which made an international star
of Bruce Lee, the Chinese–American kung-
fu artist, and launched a cycle of martial
arts movies. Lee's dynamic skills are here
employed to combat murder and mayhem
in a Bangkok ice-factory.
Rank   VHS, Beta, V2000   B

BRUCE LEE

No evil on earth
could survive the
fury that was
Bruce Lee!

All Quiet on the Western Front

The Battle of the River Plate

The Blue Max

The Big Red One
US 1980   109m   v   colour

*Lee Marvin, Mark Hamill, Robert
Carradine, Bobby DiCocco
D, Samuel Fuller*

A powerful tale, based on Fuller's own experiences with the US First Infantry Division during the Second World War, follows the fortunes of five soldiers from the North African campaign to the German surrender.
CBS/Fox   VHS, Beta, V2000, Laser   A

The Black Pirate**
US   1926   83m   □   bw

*Douglas Fairbanks, Billie Dove, Donald
Crisp, Sam de Grasse
D, Albert Parker*

First-rate Fairbanks swashbuckler in which he is a disguised nobleman avenging a pirate gang and getting his girl; whether swimming under water or sliding down the mainsail, his athleticism was rarely displayed to better effect.
Spectrum   VHS, Beta   B

The Blue Max
US   1966   149m   v   colour

*George Peppard, James Mason, Ursula
Andress, Jeremy Kemp
D, John Guillermin*

Superb battle shots and better-than-average human interest elevate the First World War story of a German pilot (Peppard) whose personal and professional ambitions bring him into conflict with his superior (Mason).
CBS/Fox   VHS, Beta, Laser   A

The Boat
W Germany   1981   138m   v   col

*Jürgen Prochnow, Herbert Grönemeyer,
Klauss Wennemann
D, Wolfgang Petersen*

The tale of a German U-boat and its crew during the Second World War. Its understated heroism and the virtual absence of a political dimension recall the British war films of the 1950s.
RCA/Columbia   VHS, Beta   R

A Bridge Too Far
US/GB   1977   175m   v   colour

*Dirk Bogarde, Michael Caine, Sean
Connery, Elliott Gould, Laurence Olivier
D, Richard Attenborough*

A painstaking and worthy account of the parachute battle of Arnhem in 1944. But the complex military operation is not always intelligible and too many stars are underused in cameo parts.
Warner   VHS, Beta   A

The Bridges at Toko-Ri*
US   1954   99m   v   colour

*William Holden, Grace Kelly, Fredric
March, Mickey Rooney
D, Mark Robson*

Holden is a jet pilot embarking on a dangerous mission during the Korean War; Rooney is the buddy who goes to his rescue and Kelly the wife waiting anxiously at home. Crisply directed study of loyalty and sacrifice.
CIC   VHS, Beta   B

Captain Blood*

US 1935 99m ☐ bw

Errol Flynn, Olivia de Havilland, Basil Rathbone, Lionel Atwill
D, Michael Curtiz

The film that made a star of Errol Flynn: he plays a British surgeon who is sold into slavery, escapes and becomes a pirate. Warner Brothers show the same panache as in their gangster films and musicals.
Warner   VHS, Beta   A

The Colditz Story

GB 1954 93m ☐ bw

John Mills, Eric Portman, Christopher Rhodes, Lionel Jeffries, Bryan Forbes
D, Guy Hamilton

An account of British prisoners of war in the maximum security prison of Colditz Castle during the Second World War. It creditably avoids schoolboy heroics and suggests that between the escape-attempts life was tedious.
Thorn EMI   VHS, Beta   A

Conan the Barbarian

US 1981 124m v colour

Arnold Schwarzenegger, James Earl Jones, Max von Sydow, Sandahl Bergman
D, John Milius

The film appears to have wider allegorical pretension. In effect, it is a rather plodding sword-and-sorcery adventure in which our hero is freed from slavery to avenge the slaying of his father and family.
Thorn EMI   VHS, Beta   A

Convoy

US 1978 106m v colour

Kris Kristofferson, Ali MacGraw, Ernest Borgnine, Burt Young
D, Sam Peckinpah

Noisy and destructive, it is relieved by flashes of humour. A trucker, Rubber Duck, has a feud with the local sheriff; and his fellow truckers rally to his support by forming a mile-long convoy along the Arizona highway.
Thorn EMI   VHS, Beta   C

Cross of Iron
GB/WG   1976   128m   ●   colour
James Coburn, James Mason,
Maximilian Schell, David Warner
D, Sam Peckinpah

Uncompromisingly harrowing account of
the experiences of a German batallion on
the Russian front in 1943, setting the
rivalries and ambitions of the soldiers
against the wider context of the brutality of
war and the agony of defeat.
Thorn EMI   VHS, Beta   A

The Cruel Sea
GB   1952   121m   □   bw
Jack Hawkins, Donald Sinden, Stanley
Baker, Denholm Elliott
D, Charles Frend

Faithful screen adaptation of Nicholas
Montsarrat's best-selling novel about a
British corvette and its crew as they do
battle with the enemy on the Atlantic high
seas during the Second World War; Hawkins authoritatively in charge.
Thorn EMI   VHS, Beta   A

The Dam Busters
GB   1954   119m   □   bw
Michael Redgrave, Richard Todd, Basil
Sydney, Derek Farr
D, Michael Anderson

The story of Dr Barnes Wallis and his
bouncing bombs which destroyed the Ruhr
dams, lifted out of the routine war epics by
Redgrave's fine portrait of Wallis as a man
who never forgets the human cost; memorable march by Eric Coates.
Thorn EMI   VHS, Beta   A

The Desert Fox*
US   1951   83m   v   bw
James Mason, Jessica Tandy, Cedric
Hardwicke, Luther Adler
D, Henry Hathaway

Brisk biopic of Field Marshal Rommel,
effectively impersonated by James Mason,
which concentrates on the later stages of
the North African campaign and his involvement in the July 1944 plot to assassinate
Hitler.
CBS/Fox   VHS, Beta, V2000   A

The Dirty Dozen*
US/Spain 1967 137m ● col
*Lee Marvin, Ernest Borgnine, Robert
Ryan, Charles Bronson, Jim Brown
D, Robert Aldrich*
Violent and bloody tale of 12 convicts
serving life sentences who are released
from jail to mount a suicide commando
mission against the Nazis; the morality may
be questionable but not Aldrich's accom-
plished direction.
MGM/UA   VHS, Beta, V2000   B

The Dogs of War
GB   1980   113m   v   colour
*Christopher Walken, Tom Berenger,
Colin Blakely, Hugh Millais
D, John Irvin*
Smooth adaptation, lacking a little in
excitement, of Frederick Forsyth's novel
about a battle-scarred mercenary (Walken)
who is approached to engineer a coup
d'état in a tottering West African state;
good location photography.
Warner   VHS, Beta, V2000   A

The Drum
GB   1938   91m   v   colour
*Sabu, Roger Livesey, Raymond Massey,
Valerie Hobson
D, Zoltan Korda*
Entertaining saga of the North-west Fron-
tier, with Sabu as an Indian prince, Massey
as the tyrant who is trying to usurp his
throne and the British caught in the middle;
Georges Périnal's colour photography is a
constant delight.
Spectrum   VHS, Beta   A

Duel*
US   1971   92m   colour
*Dennis Weaver
D, Steven Spielberg*
Macabre chase movie, originally made for
television but later a cinema cult piece, in
which a car driven by a travelling salesman
is menaced by a petrol tanker along the
backroads of California. A simple idea,
cleverly executed.
Arena   VHS, Beta   B

The Eagle Has Landed
GB   1976   118m   v   colour
*Michael Caine, Donald Sutherland,
Robert Duvall, Jenny Agutter, Donald
Pleasence
D, John Sturges*
Fair adaptation of Jack Higgins's best-
seller about a Nazi plot to capture Winston
Churchill during his visit to a Norfolk village
in 1943; both Caine as a paratroop officer
and Sutherland as an IRA man are miscast.
Precision   VHS, Beta, V2000, Laser   A

Earthquake

US 1974 116m v colour

Charlton Heston, Ava Gardner, George Kennedy, Lorne Greene, Genevieve Bujold, Walter Matthau
D, Mark Robson

The archetypal Hollywood disaster movie of the 1970s in which personal dramas are engulfed by a cataclysmic event, in this case the destruction of Los Angeles by an earthquake. A triumph for special effects.
CIC VHS, Beta A

El Cid***

US/Spain 1961 150m v colour

Charlton Heston, Sophia Loren, Raf Vallone, Genevieve Page, John Fraser
D, Anthony Mann

The Hollywood vogue in the early 1960s for big budget epics was an attempt to hit back at television by offering what TV could not: wide screen spectacle, with lavish colour, dramatic locations and large flamboyantly costumed casts. Of the films in this blockbuster cycle, *El Cid* stands out. The director, Anthony Mann, had been schooled in the Western and knew all about handling action and exploiting landscape; and he was able to draw directly on this expertise in tackling the exploits of the legendary knight Rodrigo Diaz de Bivar, alias El Cid, who drove the Moors from Spain in the eleventh century. Helped by the Australian-born cameraman, Robert Krasker, who made his reputation in the British cinema in the 1940s, Mann achieved a visual quality that is consistently outstanding, in composition, in colour, in the imaginatively chosen images of land and sea. There are many notable set pieces and the human element is not as perfunctory as usual in these epics. Loren's main contribution may be to look decorative, but Heston's El Cid is a substantial performance. The film does not look quite as stunning on the smaller screen and the video version is nearly half an hour shorter, though little that is important has been lost.
Intervision VHS, Beta, Laser A

Elephant Boy
GB 1937 76m v bw
Sabu, Walter Hudd, Allan Jeayes
D, Robert Flaherty/Zoltan Korda
The hand of the distinguished documentarist Robert Flaherty is evident in the depiction of Indian wildlife and the film is historically significant for launching the stable lad, Sabu, as an international star; but the drama has dated.
Spectrum VHS, Beta A

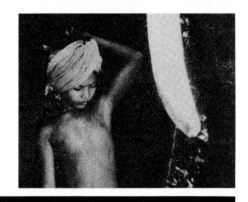

Emperor of the North
US 1973 120m v colour
Lee Marvin, Ernest Borgnine, Keith Carradine, Charles Tyner
D, Robert Aldrich

Complex, violent, three-way relationship between the king of the hobos (Marvin), a swaggering young hobo (Carradine) and a sadistic train guard (Borgnine) during the American depression of the 1930s with more than a hint of allegory.
CBS/Fox VHS, Beta, Laser A

The Enemy Below
US 1957 97m v colour
Robert Mitchum, Curt Jurgens, Theodore Bikel, David Hedison
D, Dick Powell

Duel in the South Atlantic between Mitchum's American destroyer and Jurgens's German U-boat during the Second World War. A crisply handled film which avoids making political points and tries hard to be fair to both sides.
CBS/Fox VHS, Beta, V2000, Laser A

Enter the Dragon
US/Hong Kong 1973 96m ● colour
Bruce Lee, John Saxon, Shih Kien
D, Robert Clouse

A messy attempt by Hollywood to cash in on the kung-fu craze is saved by the vitality of Lee, who plays a master of martial arts recruited by British intelligence to smash a drugs and prostitution racket.
Warner VHS, Beta A

The Fall of the Roman Empire
US/Spain 1964 178m v colour
Alec Guinness, Christopher Plummer, Stephen Boyd, James Mason
D, Anthony Mann

The unstable Commodus (Plummer) gains the throne of Rome by murdering his father (Guinness) and sees the state fall apart under his dissipated rule; more intelligently scripted than the usual Hollywood epic.
Intervision VHS, Beta A

55 Days at Peking*
US/Spain 1962 148m v colour

Charlton Heston, David Niven, Ava Gardner, Flora Robson, Robert Helpmann D, Nicholas Ray

Heston's American major, Niven's British ambassador and Gardner's Russian baroness lead the defence of the foreign legations in Peking against a fanatical religious group, the Boxers, in 1900. Production values stronger than the human interest.
Intervision   VHS, Beta   A

Fire Over England
GB   1936   91m   □   bw

Flora Robson, Laurence Olivier, Leslie Banks, Vivien Leigh, Raymond Massey D, William K Howard

Elizabeth I (Robson) leading her people to glorious victory over the Spanish Armada; strong performances from a fine cast make this an engaging flag-waver, though the battles were too obviously created in the studio tank.
Videomedia   VHS, Beta, V2000   B

The First of the Few*
GB   1942   117m   v   bw

Leslie Howard, David Niven, Rosamund John, Roland Culver D, Leslie Howard

Likeable tribute, made in the aftermath of the Battle of Britain, to the designer of the Spitfire, R J Mitchell. In its quiet patriotism and understated emotion a film very characteristic of its director and star.
Precision   VHS, Beta   A

Fist of Fury
Hong Kong 1972 100m ● colour

Bruce Lee, Noa Miao, James Tien D, Lo Wei

Kung-fu student returns to Shanghai for the funeral of his teacher, discovers that the man was murdered and starts a one-man vendetta against the killers. The production is primitive but Lee's charisma more than compensates.
Rank   VHS, Beta, V2000   A

Flying Leathernecks
US   1951   98m   v   colour

John Wayne, Robert Ryan, Janis Carter, Don Taylor D, Nicholas Ray

Wayne as a martinet marine commander leading his men into battle against the Japanese; spectacular flying sequences lift an otherwise routine war film.
Kingston   VHS, Beta   A
With: *Beyond a Reasonable Doubt* (qv)

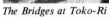

The Bridges at Toko-Ri　　　　　　　*The Dirty Dozen*

The Fall of the Roman Empire

The Four Feathers*
GB 1939 109m □ colour

*John Clements, Ralph Richardson,
C Aubrey Smith, June Duprez
D, Zoltan Korda*

The best of several screen versions of A E W Mason's famous adventure story about a young man who is branded a coward but later proves himself in the Sudan campaign of the 1890s; good location work and exciting battle scenes.
Spectrum   VHS, Beta   A

The Four Musketeers (The Revenge of Milady)
Panama/Spain 1974 102m □ col

*Oliver Reed, Raquel Welch, Richard
Chamberlain, Michael York, Frank Finlay
D, Richard Lester*

Sequel to *The Three Musketeers* (qv): the two films were originally planned as one. Not as inventive as the original but there is enough action and knockabout comedy to keep it rolling.
Thorn EMI   VHS, Beta   A

Gallipoli
Australia 1981 106m v colour

*Mark Lee, Mel Gibson, Bill Hunter,
Robert Grubb, Tim McKenzie
D, Peter Weir*

Two young men trek across the Australian desert, join the Anzac forces in Egypt and go to fight in the Dardanelles; primarily a study of comradeship set against the tragedy of war rather than a treatment of war itself.
CIC   VHS, Beta, Laser   A

Game of Death
Hong Kong 1978 95m ● colour

*Bruce Lee, Kim Tai Jong, Gig Young,
Dean Jagger, Hugh O'Brian
D, Robert Clouse*

B

A blatant attempt to exploit the Bruce Lee legend by constructing an entire film around a single fight sequence filmed by Lee shortly before his death; otherwise Lee is played by a double. A good try.
Rank   VHS, Beta, V2000   A

The Great Escape*
US 1963 170m v colour

James Garner, Steve McQueen, Richard Attenborough, Charles Bronson
D, John Sturges

The downbeat ending raises doubts about the censor's U certificate of this true story of an Allied break-out from a German prison camp. McQueen's motor cycle battle against the barbed wire is the most remembered sequence.
Warner VHS, Beta A

The Guns of Navarone*
GB 1961 157m v colour

Gregory Peck, David Niven, Stanley Baker, Anthony Quinn, Anthony Quayle
D, J Lee-Thompson

Plenty of excitement, plus some rather dubious moralizing, in this adaptation of Alistair Maclean's novel about a group of Allied commandos with a mission to destroy two huge guns during the Second World War.
RCA/Columbia VHS, Beta R

Hell Drivers
GB 1957 106m v bw

Stanley Baker, Patrick McGoohan, Herbert Lom, Peggy Cummins
D, C Raker Endfield

Silly but diverting melodrama, violent for its time, about the rivalries between lorry drivers employed by a second-rate haulage company; efficiently made and containing an early screen appearance of Sean Connery, years before James Bond.
Rank VHS, Beta B

Hell in the Pacific
US 1969 103m v colour

Lee Marvin, Toshiro Mifune
D, John Boorman

Echoes of Robinson Crusoe in this ambitious, but not entirely successful two-hander about an American pilot (Marvin) and a Japanese naval officer (Mifune) who find themselves stranded on a Pacific island during the Second World War.
Guild VHS, Beta B

Hercules Conquers Atlantis*
Italy/France 1961 80m □ col

Reg Park, Fay Spain, Ettore Manni, Luciano Marin
D, Vittorio Cottafavi

How three men and a dwarf save ancient Greece; and how a director with flair and imagination can overcome a wandering script and comic-strip heroics to produce a film of extraordinary visual power with a dazzling use of colour.
Videoform VHS, Beta B

The Heroes of Telemark
GB 1965 126m v colour
*Kirk Douglas, Richard Harris, Ulla
Jacobsson, Roy Dotrice
D, Anthony Mann*

Resistance workers set out to destroy a
heavy-water plant in occupied Norway
which is essential to the German develop-
ment of atomic fission; well-handled action
sequences and dazzling snowscapes but
also patches of tedium.
Rank VHS, Beta, V2000 A

The Hindenburg
US 1975 110m v colour
*George C Scott, Anne Bancroft, Burgess
Meredith, William Atherton
D, Robert Wise*

A late entry for the 1970s disaster cycle
which ingeniously suggests that the crash
of the German airship in New Jersey in
1937 was caused by an anti-Nazi sabo-
teur: the special effects are far more
interesting than the human drama.
CIC VHS, Beta B

Hooper
US 1978 86m v colour
*Burt Reynolds, Sally Field, Brian Keith,
Jan-Michael Vincent
D, Hal Needham*

Reynolds as an old stuntman who fears he
is past it but decides to embark on one more
big stunt before retiring; director Needham,
himself a former stuntman, keeps the
action and in-jokes flowing.
Warner VHS, Beta A

Ice Cold in Alex*
GB 1958 125m v bw
*John Mills, Sylvia Syms, Anthony
Quayle, Harry Andrews
D, J Lee-Thompson*

Suspense in plenty as an Allied medical
team in Libya during the Second World
War battles to get an ambulance to safety
through minefields, sandstorms and other
hazards; a British equivalent of *Wages of
Fear*.
Thorn EMI VHS, Beta B

In Which We Serve***
GB 1942 96m v bw
*Noel Coward, Bernard Miles, John Mills,
Celia Johnson, Kay Walsh
D, Noel Coward*

Coward's tribute to the Royal Navy, made
as a morale-booster in the dark days of the
Second World War, about a destroyer dive-
bombed in the battle of Crete; intensely
patriotic, a shade patronizing and still, in
parts, very moving.
Rank VHS, Beta A

Bruce Lee claims his revenge through death and beyond.

Fist of Fury

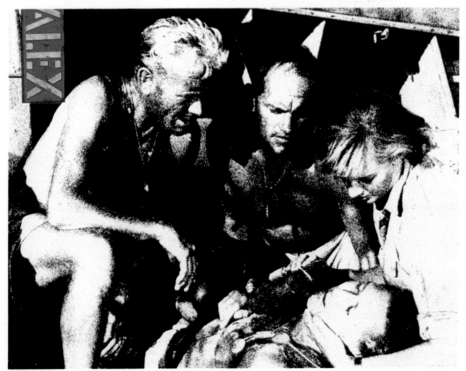

Ice Cold in Alex

Ivanhoe

GB 1952 106m v colour

*Robert Taylor, Joan Fontaine, Elizabeth
Taylor, Emlyn Williams, George Sanders
D, Richard Thorpe*

Lavish version of Sir Walter Scott's novel of
medieval England in which the valiant
knight Ivanhoe sets out on his quest for the
missing monarch, Richard the Lionheart, at
the close of the third crusade.
MGM/UA VHS, Beta B

Kagemusha***

Japan 1980 179m v colour

*Tatsura Nakadai, Tsutomu, Kenichi
Hagiwara
D, Akira Kurosawa*

Kurosawa's epic tragedy of warlords fighting for control of sixteenth-century Kyoto is
one of the modern cinema's most profound
and powerful works with a relentless
narrative surge, magnificent set pieces and
a striking use of colour.
CBS/Fox VHS, Beta, V2000, Laser A

Kelly's Heroes

US/Yugo 1970 144m v colour

*Clint Eastwood, Telly Savalas, Don
Rickles, Donald Sutherland
D, Brian G Hutton*

Echoes of *The Dirty Dozen* (qv) as Eastwood
leads his renegade US army platoon on a
bank raid in occupied France, with gold
bars as the booty. A relentless barrage of
noise and destruction, competent of its
type.
MGM/UA VHS, Beta, Laser A

Knight Without Armour**

GB 1937 96m v bw

*Robert Donat, Marlene Dietrich, Irene
Vanburgh, Herbert Lomas, Austin Trevor
D, Jacques Feyder*

Combination of Hungarian impresario
(Alexander Korda), French director, German
leading lady and English studio came up
trumps with this smashing yarn about a
widowed countess helped to safety by a
journalist during the Russian Revolution.
Spectrum VHS, Beta A

The Last American Hero**
US 1973 91m v colour

Jeff Bridges, Valerie Perrine, Geraldine Fitzgerald, Ned Beatty
D, Lamont Johnson

The alternative title, *Hard Driver*, gives a better indication of this sharply observed, often funny study of a young bootlegger's passion for stock car racing in North Carolina; writing and acting of a high class.
CBS/Fox   VHS, Beta, Laser   A

Lawrence of Arabia*
GB 1962 207m v colour

Peter O'Toole, Omar Sharif, Arthur Kennedy, Jack Hawkins, Alec Guinness
D, David Lean

A film which ambitiously attempts to do the impossible and get to the heart of one of the most enigmatic figures of the twentieth century; despite a game attempt by O'Toole, the enigma remains. Magnificent desert landscapes.
RCA/Columbia   VHS, Beta   R

The Longest Day*
US 1962 169m v bw

John Wayne, Robert Mitchum, Henry Fonda, Robert Ryan, Rod Steiger
D, Andrew Marton, Ken Annakin, Bernhard Wicki

A long, detailed and star-studded recreation of the D-Day landings in Normandy in June 1944 told from both the Allied and German points of view. A roster of familiar faces in cameo roles.
CBS/Fox   VHS, Beta, V2000, Laser   A

MacArthur
US 1977 122m v colour

Gregory Peck, Dan O'Herlihy, Ed Flanders, Ward Costello
D, Joseph Sargent

Straightforward and on the whole admiring portrait of the controversial General Douglas MacArthur from his evacuation of the Philippines in 1942 to his sacking by President Truman during the Korean War; worthy, but lacking in grandeur.
CIC   VHS, Beta   B

Mad Max

Australia 1979 88m ● colour

Mel Gibson, Joanne Samuel, Hugh Keays-Byrne, Steve Bisley
D, George Miller

Violent and destructive battle between a motor cycle gang and the police set in a nightmare future; a gruesome, unsubtle form of road movie that became a huge box-office success both in its native Australia and further afield.
Warner   VHS, Beta   A

Mad Max II

Australia 1981 91m ● colour

Mel Gibson, Bruce Spence, Vernon Wells, Emil Minty
D, George Miller

Given the success of the original, an inevitable sequel and much the same formula as before with police and bikers back in conflict; but it is arguably the better film, made with a much larger budget and greater skill and invention.
Warner   VHS, Beta   A

The Mark of Zorro

US 1920 82m ☐ bw/tint

Douglas Fairbanks, Marguerite de la Motte, Noah Beery
D, Fred Niblo

A bit creaky and stilted now, but has its place in cinema history as the film that launched Fairbanks on his dazzling career as a swashbuckling adventurer; he plays the Mexican Don Diego, a fop by day and a Robin Hood by night.
Spectrum   VHS, Beta   B

Mutiny on the Bounty

US 1962 185m v colour

Marlon Brando, Trevor Howard, Richard Harris, Hugh Griffith, Tarita
D, Lewis Milestone

Conflict on HMS Bounty, on the way to Tahiti in 1787, between Captain Bligh and his aristocratic first officer, Fletcher Christian; Brando was a controversial choice for Christian and his mannered playing will not be to all tastes.
MGM/UA   VHS, Beta   B

The Naked and the Dead

US 1958 123m v colour

Aldo Ray, Cliff Robertson, Raymond Massey, William Campbell
D, Raoul Walsh

No more than competent adaptation of Norman Mailer's novel about an American infantry platoon's mission behind enemy lines during the Pacific War.
Kingston   VHS, Beta   A
with *Target*, 1952 Western (54m)

The Night of the Generals
GB/France   1966   148m   v   col

Peter O'Toole, Omar Sharif, Tom Courtenay, Phillipe Noiret
D, Anatole Litvak

Sharif as a German intelligence officer on the trail of a Nazi general suspected of killing prostitutes during the Second World War. Makes effective use of its Warsaw and Paris locations, but it is rather long.
RCA/Columbia   VHS, Beta   A

A Night to Remember
GB   1958   123m   v   bw

Kenneth More, David McCallum, Honor Blackman, Michael Goodliffe
D, Roy Baker

A detailed, semi-documentary reconstruction of the sinking of the *Titanic*, with an intelligent script by Eric Ambler; a diffuse beginning developed into a gripping climax; strong performances.
Rank   VHS, Beta   A

North West Frontier*
GB   1959   87m   v   colour

Kenneth More, Lauren Bacall, Herbert Lom, Ursula Jeans, Wilfrid Hyde-White
D, J Lee-Thompson

A hazardous train journey across the plains of India in the early years of the century as an English colonel (More) escorts a young Hindu prince to safety; a rattling good adventure, crisply directed (abridged from cinema version).
Rank   VHS, Beta   A

Key to symbols

☐   suitable for all the family
v   vetting by parents advisable
●   adults only (mature teenagers may enjoy these films but parents should vet with care)

Note: Some films have been given a 'v' symbol although they are watched and enjoyed by a wide range of age groups (for example *Raiders of the Lost Ark* in the Action/Adventure section and the James Bond movies in Thrillers).

What the Stars mean

***   Not to be missed
**   Highly recommended
*   Well worth watching

Note: The absence of a star does not mean that a film is worthless; rather, a star denotes extra quality.

Patton: Lust For Glory***
US 1969   164m   v   colour
*George C Scott, Karl Malden, Michael
Bates, Stephen Young, Michael Strong
D, Franklin Schaffner*

The best war films have tended to be statements against war; *Patton*, on the other hand, is a portrait of a professional soldier dedicated to waging war and overcoming the enemy at almost any cost. General George S ("Blood and Guts") Patton was a strange, obsessive figure, with a total belief in his own ability and yet not above asking the Almighty for divine intervention (instructing his chaplain to pray for fine weather). The film, therefore, has plenty to bite on, as it follows Patton's campaigns in North Africa and Sicily; his famous dismissal for striking a soldier he suspected of cowardice (the man was actually suffering from battle fatigue); his partial re-instatement to lead the Third Army into Normandy; and final removal as military commander of Bavaria for being too tough on the Russians and too soft on ex-Nazis. *Patton* has its quota of spectacular battle scenes but is primarily a study of character. George C Scott, perfectly cast in the leading role, gives a warts-and-all portrayal of extraordinary depth, range and power. *Patton* won seven Oscars, including best film, best director (Schaffner) and best screenplay (Francis Ford Coppola and Edmund H. North). Scott made history by refusing to accept his Oscar, saying that he did not consider himself to be in competition with his fellow actors for rewards or recognition.

CBS/Fox   VHS, Beta, Laser   A

The Poseidon Adventure
US   1972   113m   v   colour

Gene Hackman, Ernest Borgnine, Shelley Winters, Red Buttons, Carol Lynley
D, Ronald Neame

Early disaster movie in which a luxury liner full of well-known Hollywood faces is hit by a tidal wave and capsizes; won an Oscar for visual effects.
CBS/Fox   VHS, Beta, V2000, Laser   A
also available: *Beyond The Poseidon Adventure* (Warner)

The Purple Plain
GB   1954   96m   v   colour

Gregory Peck, Maurice Denham, Lyndon Brook, Brenda de Banzie, Bernard Lee
D, Robert Parrish

Peck as a Canadian squadron leader whose plane crashes in the jungle during the Burma campaign of the Second World War; part war adventure and part psychological study of a man trying to get his nerve back.
Rank   VHS, Beta   A

Raiders of the Lost Ark*
US   1981   115m   v   colour

Harrison Ford, Karen Allen, Ronald Lacey, Paul Freeman
D, Steven Spielberg

Enormously popular pastiche of the Saturday morning serials, in which the hero, an archaeologist called Indiana Jones, finds himself on the trail of Nazi treasure hunters; noisy, incoherent at times, but has inspired moments.
CIC   VHS, Beta, Laser   C

Reach for the Sky
GB   1956   136m   v   bw

Kenneth More, Muriel Pavlow, Lyndon Brook, Lee Patterson, Alexander Knox
D, Lewis Gilbert

Second World War heroics enlivened by Kenneth More's likeable portrayal of the Battle of Britain hero, Douglas Bader, who learns to fly again after losing both legs in an accident; Pavlow plays his supportive wife.
Rank   VHS, Beta   A

Sanjuro*

Japan 1961 96m v colour

Toshiro Mifune, Tatsua Nakadai, Masao Shimuzu, Yunosuke Ito
D, Akira Kurosawa

Tifune as a scruffy, drunken old samurai warrior who has seen better days but can still show the youngsters a thing or two when it comes to dealing with a bunch of brigands. Japanese-style Western, with tongue in cheek.

Palace VHS, Beta A

The Savage Innocents

GB/France/Italy 1960 107m v col

Anthony Quinn, Yoko Tani, Marie Yang, Peter O'Toole
D, Nicholas Ray

Quinn is not convincing as an Eskimo who kills a missionary in a fit of anger and flees to the Arctic wastes with his wife. Ray intended the film as a study of cultural alienation but got bogged down; the photography is outstanding.

Rank VHS, Beta A

The Scarlet Pimpernel**

GB 1934 93m v bw

Leslie Howard, Merle Oberon, Raymond Massey, Nigel Bruce
D, Harold French

The languid Howard is perfectly cast as Sir Percy Blakeney, alias the mysterious Pimpernel rescuing aristocrats from the guillotine during the French Revolution. Unpretentious, well-made adventure that has proved surprisingly durable.

Spectrum VHS, Beta A

Scott of the Antarctic*

GB 1948 105m v colour

John Mills, James Robertson Justice, Derek Bond, Harold Warrender, Reginald Beckwith, Kenneth More
D, Charles Frend

Competent, if uninspired, account of Captain Scott's ill-fated expedition to the South Pole in 1912, with dazzling snowscapes and a distinguished score (later a symphony) by Ralph Vaughan Williams.

Thorn EMI VHS, Beta A

The Sea Hawk**
US 1940 122m ☐ bw
*Errol Flynn, Flora Robson, Brenda
Marshall, Henry Daniell
D, Michael Curtiz*

Flynn at the peak of his swashbuckling
form as a privateer wreaking havoc on the
Spanish navy under the conniving eye of
Robson's Queen Elizabeth I. A stirring yarn
of the high seas, full of furious action.
MGM/UA CED only

Sharks' Treasure
US 1974 95m ☐ col
*Cornel Wilde, Yaphet Kotto, John
Neilson, David Canary
D, Cornel Wilde*

Wilde and chums head for the shark-
infested waters of the Caribbean in search
of lost Spanish gold and find themselves
doing battle with five escaped convicts.
Old-fashioned adventure, enjoyable and
unpretentious.
Warner VHS, Beta A

The Silent Flute
US 1978 91m v colour
*Jeff Cooper, David Carradine, Roddy
McDowell, Christopher Lee, Eli Wallach
D, Richard Moore*

A latter-day martial arts saga in which
Cooper has to overcome a series of
challenges in his search for the legendary
wizard Zetan. The storyline is credited to
Bruce Lee and James Coburn; the film
lacks Lee's flair.
Rank VHS, Beta A

The Son of the Sheik*
US 1926 64m ☐ bw
*Rudolph Valentino, Vilma Banky, Agnes
Ayres
D, George Fitzmaurice*

Valentino, the "Latin lover", at the height of
his powers in what proved to be his last film
before his premature death; he plays two
parts – father and son – in a light-hearted
desert adventure crammed with fights,
chases and escapes.
Spectrum VHS, Beta B

Southern Comfort*
US 1981 99m ● colour
*Keith Carradine, Powers Boothe, Fred
Ward, Peter Coyote
D Walter Hill*

Violent, crisply handled and convincingly
acted allegory on Vietnam. A group of
National Guardsmen on a routine training
exercise in the Louisiana swamp find
themselves in a murderous battle with the
native Cajuns.
Thorn EMI VHS, Beta A

Spartacus**
US 1960 180m v colour
*Kirk Douglas, Laurence Olivier, Charles
Laughton, Jean Simmons, Peter Ustinov
D, Stanley Kubrick*

A slave revolt led by Spartacus (Douglas)
against the ancient Romans. One of the
best of the 1960s epics, thanks to intelligent scripting, a cluster of good performances (Ustinov won an Oscar) and
Kubrick's narrative control.
CIC VHS, Beta A

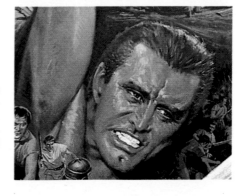

The Ten Commandments*
US 1956 219m v colour

Charlton Heston, Yul Brynner, Anne Baxter, Edward G Robinson, Yvonne de Carlo
D, Cecil B DeMille

Moses (Heston) leading the Israelites to the Promised Land in DeMille's last Biblical epic. Characteristic flair for spectacle – the parting of the Red Sea won the special effects Oscar.
CIC VHS, Beta B

The Thief of Bagdad**
US 1924 137m □ bw/tint

Douglas Fairbanks, Snitz Edwards, Charles Belcher, Julanne Johnston
D, Raoul Walsh

First-class Fairbanks vehicle in which he slays the dragon and rescues the city from the Mongols on the way to winning his princess; lavish sets and costumes by William Cameron Menzies, plus the effects which still convince.
Spectrum VHS, Beta A

The Thief of Baghdad***
GB/US 1940 100m □ colour

Conrad Veidt, Sabu, June Duprez, John Justin, Rex Ingram
D, Ludwig Berger, Michael Powell, Tim Whelan

The thief is a boy (Sabu) outwitting Veidt's evil grand vizier. One of the happiest of the Alexander Korda spectaculars, the perfect combination of pace, colour and style; the trick-photography is magical.
Spectrum VHS, Beta A

The Three Musketeers*
Panama 1973 102m □ colour

Michael York, Oliver Reed, Richard Chamberlain, Frank Finlay, Raquel Welch
D, Richard Lester

With Spike Milligan in the supporting cast and Lester (who made the Beatles films) in charge, this is a tongue-in-cheek version of the Dumas classic. Glittering costumes, dashing swordplay and a stream of gags.
Thorn EMI VHS, Beta A

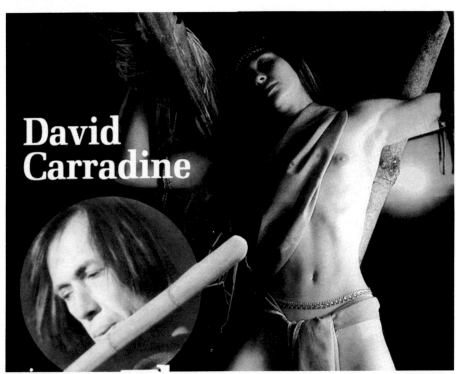

The Silent Flute

Ulysses

Too Late the Hero
US 1969 133m v colour

*Michael Caine, Cliff Robertson, Ian
Bannen, Henry Fonda, Denholm Elliott
D, Robert Aldrich*

Echoes of Aldrich's *The Dirty Dozen* in this
violent, cynical portrait of men under
stress. A bunch of unprincipled Allied
soldiers set out on a "suicide" mission
against the Japanese on a small Pacific
island during the Second World War.
Guild VHS, Beta A

Tora! Tora! Tora!
US/Japan 1970 138m v colour

*Martin Balsam, Joseph Cotten, Jason
Robards, Soh Yamamura
D, Richard Fleischer, Toshio Masuda, Kinji
Fukasaku*

Meticulous reconstruction of the events
leading up to the Japanese attack on Pearl
Harbor in December 1941, told from both
the American and Japanese points of view.
CBS/Fox VHS, Beta A

The Towering Inferno*
US 1974 165m v colour

*Paul Newman, Steve McQueen, William
Holden, Faye Dunaway, Fred Astaire
D, John Guillermin*

The fight to save hundreds trapped by fire
in the world's tallest building. Expertly
orchestrated disaster movie with all the
right ingredients: strong plot, a roster of
star names and breathtaking work by the
special effects men.
Warner VHS, Beta A

A Town Like Alice
GB 1956 115m v bw

*Virginia McKenna, Peter Finch, Maureen
Swanson, Marie Lohr
D, Jack Lee*

Realistic, episodic, studio-bound adaptation of Nevil Shute's novel about the
sufferings of a group of women taken
prisoner by the Japanese in Malaya during
the Second World War; forerunner of the
television series, *Tenko*.
Rank VHS, Beta B

The Treasure of the Sierra Madre**

US 1948 126m v bw

Humphrey Bogart, Walter Huston
D, John Huston

Three prospectors strike gold in Mexico in the 1920s but are undone by their own greed; a little stagey but strongly acted; won three Oscars for the Hustons – John (best director and screenplay), father Walter (best supporting actor).
Warner VHS, Beta A

Ulysses

Italy 1954 118m □ colour

Kirk Douglas, Silvana Mangano, Anthony Quinn, Rosanna Podesta
D, Mario Camerini

Hollywood-style treatment of ancient Greek mythology saga with more respect for its source – Homer's *Odyssey* – than is usual. Douglas as the King of Ithaca coming up against the sirens, Circe and the giant Cyclops.
Videomedia VHS, Beta, V2000 B

The Way of the Dragon*

Hong Kong 1973 88m ● colour

Bruce Lee, Nora Miao, Chuck Norris
D, Bruce Lee

An improbable story of a country boy from Hong Kong saving a Chinese restaurant threatened by gangsters in Rome. Lee's best performance also reveals him as a director of subtlety and imagination.
Rank VHS, Beta, V2000 A

We Dive at Dawn

GB 1943 97m □ bw

John Mills, Eric Portman, Reginald Purdell, Niall MacGinnis
D, Anthony Asquith

Careful and sympathetic treatment of life on a British submarine during the Second World War. Mills and Portman lead the crew of the *Sea Tiger*, making for Baltic waters with a mission to destroy the German ship, *Brandenburg*.
Rank VHS, Beta A

Where Eagles Dare

GB 1968 150m v colour

Richard Burton, Clint Eastwood, Mary Ure, Patrick Wymark, Michael Hordern
D, Brian G Hutton

Alistair MacLean's adaptation of his novel about a team of British paratroopers trying to rescue a high-ranking Allied officer from an apparently impregnable fortress in the Bavarian Alps; moves fast enough to cover the implausibilities.
MGM/UA VHS, Beta C

The Wild Geese
GB   1978   128m   v   colour
*Roger Moore, Richard Burton, Richard
Harris, Jack Watson
D, Andrew V McLaglen*

Moore, Burton, Harris and Kruger as mercenaries recruited to rescue the deposed leader of a central African state. Sound plot and action, but there are some unconvincing characters and dubious moralising.
Rank   VHS, Beta, Laser   A

The Wooden Horse*
GB   1950   98m   □   bw
*Anthony Steel, Leo Genn, David
Tomlinson, David Greene
D, Jack Lee*

Faithful reconstruction of the famous Second World War escape from Stalag Luft III in which British prisoners used a vaulting horse as cover for their tunnel. A strong story, efficiently and often excitingly told.
Thorn EMI   VHS, Beta   A

Yojimbo
Japan   1961   112m   v   bw
*Toshiro Mifune, Eijiro Tono, Seizaburo
Kawazu, Izuzu Yamada
D, Akira Kurosawa*

Kurosawa, whose *Seven Samurai* was made by Hollywood into *The Magnificent Seven*, shows his affection for the American Western with this story of a warrior who wanders into a nineteenth-century town terrorised by rival gangs.
Palace   VHS, Beta   A

SYLVIA KRISTEL is

Emmanuelle

THE ORIGINAL

X

Behind Convent Walls
Italy 1977 91m ● colour
*Ligia Branice, Marina Pierro, Gabriella
Giaccobe, Loredana Martinez
D, Walerian Borowczyk*

Sinful goings-on among sex-obsessed
nuns in a nineteenth century Italian con-
vent voyeuristically captured by Borowczyk's
camera; his lightness of touch makes this
more than just another sex film, but the
narrative is often confusing.
Canon VHS, Beta C

Bilitis
France 1976 91m ● colour
*Patricia d'Arbanville, Mona Kristensen,
Bernard Giraudeau, Gilles Kohler
D, David Hamilton*

The sexual awakening of a French schoolgirl during a summer holiday near St
Tropez told in a succession of languidly
shot, delicately hued, soft focus images
that recall director Hamilton's other career
as a fashion photographer.
Intervision VHS, Beta A

Caged Heat*
US 1974 80m ● colour
*Juanita Brown, Roberta Collins, Erica
Gavin, Barbara Steele
D, Jonathan Demme*

Deliciously overcooked melodrama about a
breakout from a sadistic women's prison
which transcends its tawdry material with
style and flair and turns a cheap exploita-
tion formula into something approaching a
feminist tract.
Iver VHS, Beta B

Cousins in Love
France/WG 1980 90m ● col
*Thierry Tevini, Jean Rougerie, Cathérine
Rouvel, Anja Shute
D, David Hamilton*

Another of David Hamilton's mistily photographed tales of young passion. A 16-year-
old French boy furthers his sexual educa-
tion against a background of approaching
war in 1939. Tastefully unerotic.
Intervision VHS, Beta A

The Decameron**
Ita/Fra/WG 1970 108m ● colour
*Franco Citti, Ninetto Davoli, Angela Luce,
Patrizia Capparelli
D, Pier Paolo Pasolini*

Eight bawdy tales from Boccaccio give
director Pasolini the opportunity for a
joyous celebration of love, lust and lechery
in the Middle Ages. Striking use of colour
and a precise feel for medieval character
and landscape.
Warner VHS, Beta A

Desperate Living
US 1977 90m ● colour
*Liz Renay, Mink Stole, Susan Lowe,
Edith Massey, Jean Hill
D, John Waters*

Waters's deliberately outrageous low-
budget movies with screeching elephan-
tine women (or drag artists) at their centre
are not to everyone's taste but they have
their cult following. This tale of rape,
murder and cannibalism is typical.
Palace VHS, Beta A

43

Emmanuelle
France   1974   90m   ● colour

Sylvia Kristel, Marika Green, Daniel Sarky, Alain Cuny
D, Just Jaeckin

Superior photography and exotic locations made this story of a French diplomat's wife and her sexual adventures in Bangkok into a landmark of softcore cinema.
Brent Walker   VHS, Beta   A
Goodbye Emmanuelle (RCA/Columbia)

Empire of Passion
France/Japan 1978 108m ● col

Kazuko Yoshiyuki, Tatsuya Fuji, Takahiro Tamura, Takuzo Kawatani
D, Nagisa Oshima

A woman is haunted by the husband she and her young lover killed. The director of the controversially erotic *Ai No Corrida* returns to the theme of obsessive impossible love, taking in elements of ghost and thriller.
Virgin   VHS, Beta   A

Female Trouble
US   1974   90m   ● colour

Divine, Edith Massey, Mink Stole, David Lochary, Mary Vivian Pearce
D, John Waters

The outsize transvestite, Divine, as a delinquent schoolgirl who leaves home when she is refused a pair of cha-cha heels, is raped by a truck driver and has a child and finishes up in the electric chair. Enough said?
Palace   VHS, Beta   A

Flesh*
US   1969   86m   ● colour

Joe Dallesandro, Geraldine Smith, Maurice Bradell, Louis Waldon
D, Paul Morrissey

A wry look at New York low life from a talented graduate of the Andy Warhol film "factory", Paul Morrissey, whose trademark is the matter-of-fact treatment of bizarre events, in this case the encounters of a young hustler.
Virgin   VHS, Beta   A

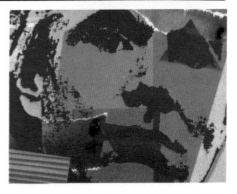

Desperate Living.

Truck Stop Women.

Vixen.

Fritz the Cat***
US 1971 78m ● colour

Director/animator, Ralph Bakshi

Had the distinction of being the first cartoon feature to be awarded an X certificate by the censor and no wonder! A scintillating farrago of sex, drugs and murder, it is wickedly funny and a scathing portrait of contemporary America.

Thorn EMI VHS, Beta A

also available: *The Nine Lives of Fritz the Cat* (Spectrum)

Heat*
US 1972 97m ● colour

Joe Dallesandro, Sylvia Miles, Andrea Feldman, Pat Ast

D, Paul Morrissey

Young man arrives in Hollywood hoping to become a star but finds himself involved with the sort of sexual and social misfits without which no Warhol/Morrissey film would be complete. Rude, crude and frequently funny.

Virgin VHS, Beta A

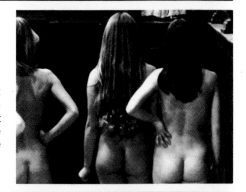

Immoral Tales
France 1974 98m ● colour

Lise Danvers, Charlotte Alexandra, Paloma Picasso, Florence Bellamy

D, Walerian Borowczyk

Four bawdy stories, starting in the present day and working back to Lucrezia Borgia in 1498. Much dwelling on naked flesh but lifted out of the rut of the average sex movie by the fresh, poetic quality of many of the images.

Thorn EMI VHS, Beta A

Lonesome Cowboys
US 1968 90m ● colour

Taylor Mead, Louis Waldon, Viva, Eric Emerson

D, Andy Warhol

A Western by Warhol was bound to be like no other and so it proved, with the men all homosexual and the sole female doing her best to get herself raped. Played relentlessly for laughs, but a little too self-conscious for its own good.

Virgin VHS, Beta A

Multiple Maniacs
US   1970   90m   ●   bw
Divine
D, John Waters

The appalling Divine, complete with grotesque blonde wig, plays the owner of a freak show who lures her curious customers to their death. With *Cocaine Fiends* (1939), a Hollywood B film so dreadful as to be almost funny.
Palace   VHS, Beta   A

Three Immoral Women
France   1978   89m   ●   colour
Marina Pierro, Gaëlle Legrand, Pascale Christophe
D, Walerian Borowczyk

Another sex portmanteau from Borowczyk, including a story about the infatuation of the painter Raphael with a peasant girl. Like all the director's films, this is often stunning to look at but otherwise there is less to recommend it.
Walton   VHS, Beta   A

Trash*
US   1970   95m   ●   colour
Joe Dallesandro, Holly Woodlawn, Jane Forth, Michael Sklar, Geri Miller
D, Paul Morrissey

One of the best Morrissey/Warhol efforts. A young man is so emotionally deadened by his addiction to drugs that he is incapable of communicating with those around him and instead becomes a neutral focus for their fantasies.
Virgin   VHS, Beta   A

Truck Stop Women
US   1974   88m   ●   colour
Lieux Dressler, Claudia Jennings, Gene Drew, Dolores Dorn, Dennis Fimple
D, Mark L Lester

A joyous romp through B movie sex-and-gangster territory which, like *Caged Heat*, comes out a lot better than its material. Anna's Truck Stop in the American southwest is both a source of girls and cover for a hijack racket.
Inter-Ocean   VHS, Beta, V2000   C

Vixen
US   1968   85m   ●   colour
Erica Gavin, Harrison Page, Garth Pillsbury, Michael Donovan O'Donnell, Vincene Wallace
D, Russ Meyer

Sex film with tongue firmly in cheek as Meyer mixes the obligatory couplings with a sub-plot about hijacking a plane to Cuba; the sheer absurdity, plus spirited performances, just about carry the film through.
Videospace   VHS, Beta   A

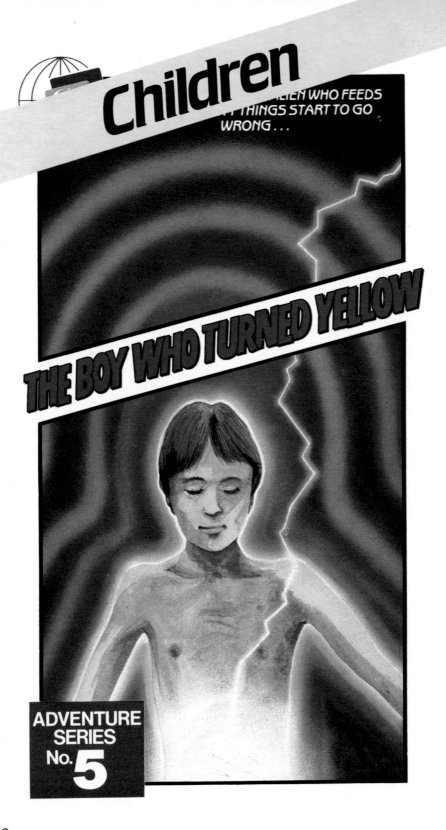

Children

...LIEN WHO FEEDS
...T THINGS START TO GO
WRONG . . .

THE BOY WHO TURNED YELLOW

ADVENTURE
SERIES
No. 5

The Adventures of Tom Sawyer**
US 1938 76m □ colour

Tommy Kelly, Ann Gillis, May Robson,
Victor Jory, Walter Brennan
D, Norman Taurog

Faithful adaptation of the classic Mark Twain story about the boy from the banks of the Mississippi who tracks down the murderer, Injun Joe. Attractive photography by James Wong Howe.
Guild VHS, Beta A

The Apple Dumpling Gang
US 1974 100m □ colour

Bill Bixby, Susan Clark, Don Knotts, Tim Conway, David Wayne
D, Norman Tokar

Three orphan children find gold in California in 1878 and have to stop it falling into the hands of their evil guardian. Lively and enjoyable adventure comedy with inventive gags and nicely judged characterisations.
Disney VHS, Beta R

Bedknobs and Broomsticks
US 1971 97m □ colour

Angela Lansbury, David Tomlinson,
Roddy McDowall, Sam Jaffe
D, Robert Stevenson

Three cockney children evacuated from London to the coast in 1940 are put in charge of a kindly witch and together they are able to thwart a German invasion. A follow-up to *Mary Poppins* (qv), with the same mixture of live action and animation.
Disney VHS, Beta R

Black Beauty
GB 1971 104m □ colour

Mark Lester, Walter Slezak, Peter Lee Lawrence, Patrick Mower
D, James Hill

Appealing version, faithful to the book, of Anna Sewell's classic story of a horse which has a series of adventures while passing through several hands before being returned to its original young owner. The animal is the star.
Video Movies VHS, Beta A

Blackbeard's Ghost
US 1967 104m ☐ colour

Peter Ustinov, Dean Jones, Suzanne Pleshette, Elsa Lanchester
D, Robert Stevenson

Blackbeard the pirate comes back as a ghost to help a group of old ladies who are in financial trouble with their inn because a crook wants to turn it into a casino. Ustinov's irrepressible ghost is a joy.
Disney   VHS, Beta   R

The Boatniks
US 1970 95m ☐ colour

Robert Morse, Stephanie Powers, Phil Silvers, Norman Fell, Mickey Shaughnessey
D, Norman Tokar

Accident-prone coast guard has the chance to redeem himself when he comes up against a gang of jewel thieves, led by the splendid Phil Silvers. Enjoyable comedy despite too much talk.
Disney   VHS, Beta   R

The Boy Who Turned Yellow
GB 1972 55m ☐ colour

Mark Dightam, Robert Eddison, Brian Worth, Helen Weir, Esmond Knight
D, Michael Powell

A schoolboy sent home for dozing off in class is startled to realize that he, his fellow passengers and the tube train in which they are travelling have turned yellow. First children's film by one of Britain's finest directors.
Rank   VHS, Beta   C

Candleshoe
US 1977 101m v colour

David Niven, Helen Hayes, Jodie Foster, Leo McKern, Veronica Quilligan
D, Norman Tokar

American orphan Jodie trying to pass herself off as the long-lost heiress to an English stately home in a plot hatched by the devious McKern; but butler Niven gets suspicious. Comedy-thriller with furious climax.
Disney   VHS, Beta   R

Chitty Chitty Bang Bang*
GB 1968 145m ☐ colour

Dick Van Dyke, Sally Ann Howes, Lionel Jeffries, Robert Helpmann, Gert Fröbe
D, Ken Hughes

Van Dyke as a scatter-brained inventor who saves a derelict racing car and gives it magical powers which are used against a wicked baron. Children lap it up, although adults might find it an uneasy mixture of sentimentality and farce.
Warner   VHS, Beta   A

Davy Crockett
US 1955 88m ☐ colour

Fess Parker, Buddy Ebsen, Basil Ruysdael, William Bakewell
D, Norman Foster

Appealing tale of the Tennessee hunter in the coonskin cap, as he wrestles with bears, becomes an Indian scout, enters Congress and leads the heroic fight for Texan independence at the battle of the Alamo.
Disney   VHS, Beta   R

The Devil and Max Devlin
US   1981   96m   v   colour

*Elliott Gould, Bill Cosby, Susan Anspach,
Adam Rich*
D, Steven Hilliard Stern

Max is given a chance to save himself from
Hell if he can persuade three children to
sell their souls to the devil. Not as black a
comedy as it sounds, but lacks style and
coherence.
Disney   VHS, Beta   R

Dr Who and the Daleks
GB   1965   78m   v   colour

*Peter Cushing, Roy Castle, Jennie
Linden, Roberta Tovey*
D, Gordon Flemyng

Early spin-off from the television series has
Cushing as the kindly doctor taking on the
villainous robotic Daleks with his grand-
daughters. Fair fare for Dr Who fans.
Thorn EMI   VHS, Beta   A
also available: *Daleks: Invasion Earth 2150
AD* (1966), Thorn EMI.

Dragonslayer
US   1981   106m   v   colour

*Peter MacNichol, Caitlin Clarke, Ralph
Richardson, John Hallam*
D, Matthew Robbins

Young sorcerer's apprentice uses his master's magic amulet to do battle with the
dragon that has been terrifying the kingdom. The dragon is convincing, but the film
is short on imagination and lacks narrative
clarity.
Disney   VHS, Beta   R

Escape to Witch Mountain
US   1974   91m   v   colour

*Eddie Albert, Ray Milland, Donald
Pleasence, Kim Richards*
D, John Hough

Two orphan children with supernatural
powers come under the wing of an
unscrupulous multi-millionaire (nicely
played by Milland) who hopes to exploit
their gifts for his own ends. Strong story
sometimes undermined by weak jokes.
Disney   VHS, Beta   R

Freaky Friday
US   1976   94m   □   colour

*Jodie Foster, Barbara Harris, John Astin,
Patsy Kelly*
D, Gary Nelson

A 13-year-old girl and her mother try to put
an end to their constant bickering by
changing places; but this brings new
problems. A splendid comic idea developed
with limited flair and at too great length.
Disney   VHS, Beta   R

Blackbeard's Ghost

Davy Crockett

Escape to Witch Mountain

Friend or Foe*
GB 1981 70m □ colour

*Mark Luxford, John Holmes, Stacey
Tendetter, John Bardon, Jennifer Piercey
D, John Krish*

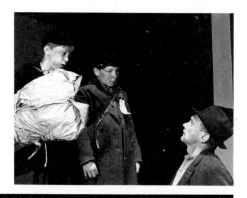

Award-winning adventure from the Children's Film and Television Foundation about two boys looking for a crashed plane in wartime Britain. One of the boys falls into a river and is saved from drowning ... by the plane's pilot.
Rank   VHS, Beta   C

The Glitterball*
GB 1977 56m □ colour

*Ben Buckton, Keith Jayne, Ron Pember,
Marjorie Yates, Barry Jackson
D, Harley Cockliss*

Fast-moving and enjoyable comic fantasy, with excellent special effects, about a friendly space traveller with an enormous appetite for human food. Won prizes at the Moscow and Los Angeles film festivals.
Rank   VHS, Beta   C

The Great Muppet Caper*
GB 1981 98m □ colour

*The Muppets, Diana Rigg, Robert Morley,
Trevor Howard, Peter Ustinov
D, Jim Henson*

More successful attempt than *The Muppet Movie* (qv) at translating the famous puppet series from television to cinema. Kermit the Frog, Fozzie Bear and Gonzo are investigative journalists following up a jewel robbery.
Precision   VHS, Beta, V2000, Laser   A

The Incredible Journey*
US 1963 80m □ colour

*Emile Genest, John Drainie, Tommy
Tweed, Sandra Scott
D, Fletcher Markle*

Three animals – a bull terrier, a golden retriever and a Siamese cat – make a 250-mile trek across Canada to be reunited with their owners. The four-legged stars do not let the film down.
Disney   VHS, Beta   R

Jason and the Argonauts*
GB 1963 102m ☐ colour

Todd Armstrong, Honor Blackman, Niall MacGinnis, Andrew Faulds
D, Don Chaffey

Jason and his shipmates seek the Golden Fleece and are beset by hazards. Handled with a nice light touch; there is excellent trick work from a master craftsman, Ray Harryhausen.
RCA/Columbia VHS, Beta A

Little Lord Fauntleroy
GB 1980 90m ☐ colour

Ricky Schroder, Alec Guinness, Eric Porter, Colin Blakely, Connie Booth
D, Jack Gold

Routine treatment of Frances Hodgson Burnett's novel about the little American boy, Ceddie Errol, who discovers that he is the heir to an English title. But the strong, old-fashioned story may be enough.
Thorn EMI VHS, Beta A

The Love Bug***
US 1968 104m ☐ colour

Dean Jones, Michele Lee, David Tomlinson, Buddy Hackett
D, Robert Stevenson

Unsuccessful racing driver buys a little Volkswagen car, finds it has a mind and heart of its own, and outwits the rascally dealer who tries to recover it. Fast-moving and very funny comedy.
Disney VHS, Beta R
also available: *Herbie Rides Again* (Disney)

Mary Poppins***
US 1964 134m ☐ colour

Julie Andrews, Dick Van Dyke, David Tomlinson, Glynis Johns, Hermione Baddeley
D, Robert Stevenson

Julie Andrews in a triumphant film debut won an Oscar as the magical Edwardian nanny who floats in on her umbrella to take charge of two naughty children. Some animation and several catchy songs.
Disney VHS, Beta R

The Muppet Movie
GB 1979 94m ☐ colour

The Muppets, Elliott Gould, Bob Hope,
Mel Brooks, Orson Welles
D, James Frawley

The adventures of Kermit the Frog and company as they travel across the United States to seek fame in Hollywood. First, and only partly successful, big screen treatment of the popular television show.
Precision   VHS, Beta, V2000, Laser   A

Never a Dull Moment*
US 1967 87m ☐ colour

Dick Van Dyke, Edward G Robinson,
Dorothy Provine, Henry Silva
D, Jerry Paris

Van Dyke in good form as a television actor who is forced to impersonate a hoodlum to get out of a nasty spot; strong support from Robinson as a crooked art dealer.
Disney   VHS, Beta   R

Old Yeller*
US 1957 83m ☐ colour

Dorothy McGuire, Fess Parker, Tommy
Kirk, Kevin Corcoran
D, Robert Stevenson

A stray mongrel adopted by a boy and his family in a remote part of Texas protects them from bears and wolves. A sentimental story for animal lovers; good of its kind.
Disney   VHS, Beta   R

One of Our Dinosaurs Is Missing*
US 1975 90m ☐ colour

Helen Hayes, Peter Ustinov, Derek
Nimmo, Clive Revill, Joan Sims
D, Robert Stevenson

Microfilm with a secret formula, smuggled out of China, is hidden in the skeleton of a dinosaur; but a Chinese intelligence officer, deliciously played by Ustinov, is soon on the trail. Enjoyable comedy.
Disney   VHS, Beta   R

The Railway Children***
GB 1970 104m □ colour
*Dinah Sheridan, Bernard Cribbins,
William Mervyn, Jenny Agutter, Sally
Thomsett
D, Lionel Jeffries*
Excellent adaptation of E Nesbit's classic
about the adventures befalling three chil-
dren and their mother in Yorkshire while
their father is in prison. Good location work.
Thorn EMI VHS, Beta A

Return from Witch Mountain
US 1978 89m □ colour
*Kim Richards, Ike Eisenmann, Bette
Davis, Christopher Lee, Jack Soo
D, John Hough*
Sequel to *Escape from Witch Mountain* (qv)
with Lee, as a mad scientist, and Davis
trying to manipulate cute orphan kids with
supernatural powers. Smoothly directed
with good special effects.
Disney VHS, Beta R

Sky Pirates
GB 1976 60m □ colour
*Adam Richens, Michael McVey, Sylvia
O'Donnell, Bill Maynard, Reginald Marsh
D, Pennington Richards*
Clean-cut goodies v baddies adventure.
Young aircraft enthusiasts hatch a clever
plan to stop crooks using a radio-controlled
plane to smuggle a stolen diamond across
the English Channel.
Rank VHS, Beta C

Snowball Express
US 1972 92m □ colour
*Dean Jones, Nancy Olson, Henry
Morgan, Kennan Wynn, Mary Wickens
D, Norman Tokar*
Fun in the snow as a New York family
inherits a run-down hotel in the Rockies
and decides to turn it into a ski resort. But
the local banker wants the property himself
and refuses to advance a loan for vital
repairs.
Disney VHS, Beta R

Swallows and Amazons

GB 1974 88m ☐ colour

Virginia McKenna, Ronald Fraser, Simon West, Sophie Neville
D, Claude Whatham

Straightforward adaptation of Arthur Ransome's story of the friendly rivalry between two groups of children on holiday in the Lake District. Attractive photography, a feel for period (the 1920s) and Fraser splendid as Uncle Jim.

Thorn EMI   VHS, Beta   A

The Swiss Family Robinson*

GB 1960 126m ☐ colour

John Mills, Dorothy McGuire, James MacArthur, Tommy Kirk, Sessue Hayakawa
D, Ken Annakin

The desert island adventures of a shipwrecked family, who explore the unknown and encounter a bunch of of pirates. A pleasant rendering of the famous children's story with Caribbean locations.

Disney   VHS, Beta   R

The Tales of Beatrix Potter**

GB 1971 86m ☐ colour

Carole Ainsworth, Sally Ashby, Frederick Ashton, Avril Bergen, Michael Coleman
D, Reginald Mills

The animal characters of Beatrix Potter played and danced by members of the Royal Ballet in adaptations of five of the tales. The animal masks are splendidly lifelike and the film makes charming use of Lake District settings.

Thorn EMI   VHS, Beta   B

That Darn Cat!*

US 1965 112m ☐ colour

Hayley Mills, Dean Jones, Dorothy Provine, Roddy McDowall, Neville Brand
D, Robert Stevenson

Lively animal comedy in which a Siamese cat helps the FBI to free a woman held hostage by a couple of bank robbers. Though the tone is light-hearted, there is effective suspense as well; plus strong character parts.

Disney   VHS, Beta   R

Tightrope to Terror

GB 1983 53m □ colour

Richard Owens, Rebecca Lacey, Eloise
Ritchie, Stuart Wilde, Mark Jefferis
D, Robert Kellett

Danger and excitement for four children
left dangling in a damaged cable car high
above a glacier. Well handled adventure
with a veiled message about not setting off
into the mountains without proper equip-
ment.
Rank VHS, Beta C

The Water Babies

GB/Poland 1978 83m □ colour

James Mason, Billie Whitelaw, Bernard
Cribbins, Joan Greenwood
D, Lionel Jeffries

A charming treatment – part-live, part-
animation – of Charles Kingsley's novel
about the underwater adventures of a 12-
year-old chimney sweep in Victorian
London. Rich performances from a well-
chosen cast.
Thorn EMI VHS, Beta A

With Children in Mind

Other films suitable for children. See
also the Bond films listed on page 282.

Action and Adventure
The Adventures of Robin Hood***
The Black Pirate**
Captain Blood*
The Four Feathers*
Raiders of the Lost Ark*
The Thief of Baghdad (1940)***
The Three Musketeers*

Comedy
Charlie Chaplin I**, II**, III**, IV**
City Lights ***
The General ***
The Gold Rush***
The Great Race*
It's a Mad, Mad, Mad, Mad World*
The Lavender Hill Mob***
Monsieur Hulot's Holiday***
On the Beat/Trouble in Store
The Pink Panther*
Those Magnificent Men in their Flying
   Machines*
Whisky Galore***

Drama
Animal Farm
Chariots of Fire***
Oliver Twist***
Popeye
The Red Shoes***
Watership Down*

Musicals
Annie*
Bugsy Malone**
Dr Dolittle
Oliver*
The Sound of Music***

Science Fiction/Fantasy
Flash Gordon*
Lord of the Rings
Star Wars**
Superman*
Superman II*
Superman III*
20,000 Leagues Under the Sea*
Voyage to the Bottom of the Sea

Westerns
Shane***

The Wizard of Oz***

US 1939 98m □ colour

Judy Garland, Frank Morgan, Ray Bolger,
Jack Haley, Bert Lahr
D, Victor Fleming

One of the most successful films ever made for children (barring Disney cartoons), it is among the few pictures of its time that can still be enjoyed by all ages. The reasons are not hard to find. There is L Frank Baum's enchanting story, faithfully adapted for the screen, about Dorothy, the Kansas farm girl, who is knocked unconscious in a tornado and finds herself setting off with her three strange friends in search of the Emerald City and the Wizard. There is the young Judy Garland as Dorothy, with her appealing wide-eyed innocence; strong character playing from Bolger (the Scarecrow), Haley (the Tin Man) and Lahr (the

Cowardly Lion), with Margaret Hamilton's Wicked Witch not far behind. There are the rich Technicolor settings (the opening and closing sequences sandwiching Dorothy's adventure, are in black and white). There are the unforgettable songs: "We're off to See the Wizard", "Follow the Yellow Brick Road" and the enduringly poignant "Over the Rainbow". It is interesting to reflect on how the film would have turned out had Shirley Temple played the lead, as MGM wanted; or if the original casting of Bolger as the Tin Man and Buddy Ebsen as the Scarecrow had gone ahead. Though Fleming is the credited director, he left before the end to take over *Gone With the Wind* and some of the sequences were shot by another distinguished Hollywood hand, King Vidor.

MGM/UA VHS, Beta, CED C

THORN EMI

HAND MADE FILMS

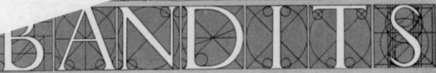

Comedy

TIME BANDITS

All the dreams you've ever had–and not just the good ones...

JOHN CLEESE · SEAN CONNERY
SHELLEY DUVALL · KATHERINE HELMOND
IAN HOLM · MICHAEL PALIN
RALPH RICHARDSON · PETER VAUGHAN
DAVID WARNER

Produced and Directed by
TERRY GILLIAM

Abbott and Costello Meet Captain Kidd
US 1952 68m □ colour

*Bud Abbott, Lou Costello, Charles
Laughton, Hilary Brooke, Leif Erickson
D, Charles Lamont*

The once incredibly popular pair in a feeble
pirate spoof. But it is the only Abbott and
Costello film on video and worth watching
for Laughton's bizarre stoogery. With: *The
Three Musketeers* a 1935 adventure.
Kingston   VHS, Beta   A

The Adventure of Sherlock Holmes' Smarter Brother
US 1975 87m v colour

*Gene Wilder, Madeline Kahn, Marty
Feldman, Dom DeLuise, Leo McKern
D, Gene Wilder*

Attempt by director/writer/star Wilder to
have fun at the expense of Sherlock
Holmes by turning over his cases to a
younger brother; an uneasy mix of bawd
and farce with inspired moments.
CBS/Fox   VHS, Beta, Laser   A

Airplane*
US 1980 87m v colour

*Robert Stack, Lloyd Bridges, Robert Hays,
Julie Hagerty, Peter Graves
D, Jim Abrahams*

Lampoon of the disaster movie in which an
ex-pilot is forced to take over the controls
and make an emergency landing when the
crew goes down with food poisoning.
Unsubtle but often hilarious.
CIC   VHS, Beta   C
Also available: *Airplane II* (CIC)

Alice's Restaurant*
US 1969 108m ● colour

*Arlo Guthrie, Pat Quinn, James
Broderick, Michael McClanathan
D, Arthur Penn*

A wry, funny and uncondescending look at
the American youth protest movement of
the 1960s with Arlo Guthrie virtually
playing himself, a young folk singer taking
to the road and doing his thing as the
Vietnam draft looms.
MGM/UA   VHS, Beta   A

Annie Hall***
US 1977 93m ● colour

*Woody Allen, Diane Keaton, Tony
Roberts, Carol Kane, Paul Simon
D, Woody Allen*

A "nervous romance" between a Jewish
comedian and an aspiring singer, based on
Allen's real-life relationship with Diane
Keaton; told with humour and perception
and avoiding the self-indulgent wallow of
some other Allen films.
Warner   VHS, Beta   A

Abbott and Costello meet Captain Kidd   *Arsenic and Old Lace*

Avanti!

Arsenic and Old Lace*
US 1944 118m v bw

*Cary Grant, Josephine Hull, Jean Adair,
Raymond Massey, Peter Lorre
D, Frank Capra*

Riotously funny version of Joseph Kesselring's black comedy about the two sweet old ladies who poison their guests with elderberry wine and get their mad brother to bury the bodies. Grant as the nephew who spoils the fun.
Warner CED only C

Arthur
US 1981 97m ● colour

*Dudley Moore, John Gielgud, Geraldine Fitzgerald, Liza Minnelli
D, Steve Gordon*

Moore as a drunken layabout who, to inherit the family fortune, must mend his behaviour and forsake the girl he loves for a marriage of convenience. Gielgud, as the deadpan butler, won an Oscar.
Warner VHS, Beta A

Avanti!*
US 1972 140m ● colour

*Jack Lemmon, Juliet Mills, Clive Revill,
Edward Andrews
D, Billy Wilder*

Harassed American executive and young Englishwoman find romance on Ischia where his father and her mother had been lovers before them. Comedy with a blackish hue and somewhat long, but infused with characteristic Wilder wit.
Warner VHS, Beta A

Bananas
US 1971 82m v colour

*Woody Allen, Louise Lasser, Carlos Montalban, Jacobo Morales
D, Woody Allen*

Bored with his job as a product-tester for a New York corporation, Woody looks for excitement in a Latin American state and accidentally becomes its dictator. More a succession of gags than a coherent whole.
Warner VHS, Beta A

Barefoot in the Park
US 1967 102m v colour

*Robert Redford, Jane Fonda, Mildred Natwick, Charles Boyer
D, Gene Saks*

Pleasant version of Neil Simon's stage hit about newlyweds who move into a run-down, top-storey, Greenwich Village apartment; and also have to cope with the bride's mother.
CIC VHS, Beta B

Being There*
US 1979 121m v colour

*Peter Sellers, Shirley MacLaine, Melvyn Douglas, Jack Warden, Richard Dysart
D, Hal Ashby*

Gentle fable about an illiterate gardener (Sellers, in his penultimate screen role) who is brought out of seclusion and through his homespun wit becomes a national celebrity. A likeable film, unsure of its ultimate purpose.
CBS/Fox VHS, Beta, V2000, Laser A

Best Friends*
US 1982 105m v colour

*Burt Reynolds, Goldie Hawn, Jessica
Tandy, Barnard Hughes
D, Norman Jewison*

Romantic comedy about two Hollywood
scriptwriters who after living happily together decide to get married and find
themselves funning into problems. Incon-
sequential but with some neat touches and
excellent central performances.
Warner   VHS, Beta   A

Billy Liar**
GB 1963 94m v bw

*Tom Courtenay, Wilfred Pickles, Mona
Washbourne, Julie Christie
D, John Schlesinger*

The novel and play by Keith Waterhouse
and Willis Hall was slightly coarsened for
the cinema. An undertaker's clerk leads a
fantasy life to escape his drab surroundings and nagging family. Courtenay plays
him to a tee.
Thorn EMI   VHS, Beta   C

Blazing Saddles*
US 1974 89m v colour

*Cleavon Little, Gene Wilder, Slim
Pickens, Harvey Korman, Madeline Kahn
D, Mel Brooks*

Joke-a-minute Wild West lampoon in
which a black sheriff and alcoholic ex-
gunfighter clean up the town of Rock Ridge.
Kahn does a brilliant Marlene Dietrich and
there are plenty of digs at other Westerns.
Warner   VHS, Beta   A

The Blues Brothers
US 1980 133m v colour

*John Belushi, Dan Aykroyd, Kathleen
Freeman, James Brown, Henry Gibson
D, John Landis*

Two brothers try to raise the money to save
their old orphanage by reviving their
rhythm 'n' blues band and leave chaos in
their wake. A noisy, destructive and self-
indulgent film with a big car chase.
CIC   VHS, Beta   A

Bob and Carol and Ted and Alice*
US 1969 105m ● colour

*Natalie Wood, Robert Culp, Elliott Gould,
Dyan Cannon
D, Paul Mazursky*

Two Californian couples decide after
therapy to own up to their extra marital
affairs and arrange a partner-swopping
orgy in Las Vegas. Fashionable and slightly
daring for its time.
RCA/Columbia   VHS, Beta   A

Cactus Flower
US 1969 104m v colour

Walter Matthau, Ingrid Bergman, Goldie Hawn, Jack Weston, Rick Lenz
D, Gene Saks

Bachelor dentist Matthau gets his secretary (Bergman) to pose as his wife in order to make things right with his young mistress (Hawn). Agreeable comedy of misunderstandings, with skilful performances by all concerned.
RCA/Col   VHS, Beta   A

La Cage aux Folles
France/Italy 1978 91m ● colour

Ugo Tognazzi, Michel Serrault, Michel Galabru, Claire Maurier
D, Edouard Molinaro

A young man about to be married tries to conceal from his moralistic future parents-in-law that his father is a homosexual living with a drag queen. Well-received, but promises more than it delivers.
Warner   VHS, Beta   A

California Suite
US 1978 99m v colour

Maggie Smith, Michael Caine, Walter Matthau, Elaine May, Jane Fonda
D, Herbert Ross

Four separate adventures of guests staying at the Beverly Hills Hotel during Academy Awards week, adapted by Neil Simon from his play. The best involves Maggie Smith nominated for an Oscar, which this performance later won.
RCA/Columbia   VHS, Beta, CED   A

The Captain's Table
GB 1958 86m v colour

John Gregson, Peggy Cummins, Donald Sinden, Reginald Beckwith, Nadia Gray
D, Jack Lee

Gregson as the captain of a cargo boat who is given the chance to command a luxury liner. Undemanding, with predictable jokes, from a novel by Richard Gordon (of the "Doctor" series).
Rank   VHS, Beta   B

The Card*
GB 1952 90m v bw

Alec Guinness, Glynis Johns, Petula Clark, Valerie Hobson
D, Ronald Neame

Guinness felicitously cast as Arnold Bennett's Denry Machin, the lad from the potteries determined to climb his way up the social ladder by fair means and foul and becoming his town's youngest mayor.
Rank   VHS, Beta   A

The "Carry On" series

All D, Gerald Thomas

Started in 1958 with *Carry on Sergeant* and continued for 20 years, becoming a popular art form the potency of which cannot be denied. Purists found the humour crude and obvious but the public kept watching. The early films relied mainly on broad slapstick but as censorship relaxed during the 1960s, the jokes became increasingly based on sexual innuendo. The last of the 29, *Carry On Emmanhuelle*, was a spoof on a celebrated softcore sex film. What redeemed the smut was the spirited playing of a gifted team of comedy actors, including Kenneth Williams, Kenneth Connor, Charles Hawtrey, Sid James, Bernard Bresslaw, Joan Sims, Barbara Windsor, Peter Butterworth and Hattie Jacques.

Titles available:

Carry on Abroad
1972   89m   v   colour
Rank   VHS, Beta
Carry On At Your Convenience
1971   88m   v   colour
Rank   VHS, Beta
Carry On Behind
1975   88m   v   colour
Rank   VHS, Beta
Carry On Camping
1968   86m   v   colour
Rank   VHS, Beta
Carry On Cleo
1964   91m   v   colour
Thorn EMI   VHS, Beta

Carry on Follow That Camel
1965   94m   v   colour
Rank   VHS, Beta
Carry On Cowboy
1966   91m   □   colour
Thorn EMI   VHS, Beta
Carry On Dick
1974   89m   v   colour
Rank   VHS, Beta
Carry On Doctor
1968   94m   v   colour
Rank   VHS, Beta
Carry On Emmannuelle
1978   90m   v   colour
VCL   VHS, Beta
Carry On England
1976   89m   v   colour
Rank   VHS, Beta
Carry On Girls
1973   88m   v   colour
Rank   VHS, Beta
Carry On Henry
1971   88m   v   colour
Rank   VHS, Beta
Carry On Matron
1972   108m   □   colour
Rank   VHS, Beta
Carry On Nurse
1959   84m   □   bw
Thorn EMI   VHS, Beta
Carry On Up The Jungle
1970   87m   v   colour
Rank   VHS, Beta
Carry On Up The Khyber
1968   86m   v   colour
Rank   VHS, Beta

Cat Ballou**
US 1965 104m v colour
*Jane Fonda, Lee Marvin, Michael Callan,
Dwayne Hickman, Nat King Cole
D, Elliot Silverstein*

Lively spoof Western hits more often than it misses. Marvin plays a drunken gunfighter and the silver-nosed villain – and won an Oscar; Fonda is a teacher turned outlaw to avenge her father's death.
RCA/Columbia   VHS, Beta, CED   A

Catch 22*
US 1970 122m v colour
*Alan Arkin, Jon Voight, Art Garfunkel,
Buck Henry, Martin Balsam
D, Mike Nichols*

Arkin superb as the man who tries to get excused bombing missions by being declared insane and is told no sane man would want to fly; otherwise a patchily successful translation of Joseph Heller's gruesome Second World War satire.
CIC   VHS, Beta   A

Charlie Chaplin I**
US 1915/17 50m □ bw
*Charles Chaplin, Edna Purviance, Henry
Bergman, Eric Campbell
D, Charles Chaplin*

Three early two-reelers: *The Tramp* (1915), *The Pawnshop* (1916) and *Easy Street* (1917). The second contains the famous sequence of Charlie taking a watch to pieces; in the third he turns policeman to overcome the local bully.
Spectrum   VHS, Beta   B

Charlie Chaplin II**
US 1915/17 111m □ bw
*Charles Chaplin, Edna Purviance, Eric
Campbell, Ben Turpin, Leo White
D, Charles Chaplin*

A further six vintage two-reelers demonstrating Chaplin's unique mixture of comedy and pathos: *The Immigrant* (1917), *His New Job* (1915), *The Vagabond* (1916), *The Champion* (1915), *Work* (1915), and *The Adventurer* (1917).
Spectrum   VHS, Beta   A

Charlie Chaplin III**
US 1914/19 115m □ bw
*Charles Chaplin, Edna Purviance, Eric
Campbell, Lloyd Bacon, Leo White
D, Charles Chaplin*

Six lesser known two-reelers: *A Woman* (1915), *The Floorwalker* (1916), *The Fireman* (1916), *The Count* (1916), *The Rink* (1916) and *The Jitney Elopement* (1915); plus the 1914 one reeler, *Making a Living*.
Spectrum   VHS, Beta   A

Charlie Chaplin IV**
US 1914/16 54m □ bw
*Charles Chaplin, Fatty Arbuckle, Minta
Dufee, Ben Turpin, Edna Purviance
D, Charles Chaplin*

Includes the 1916 two-reeler, *One A.M.*, in which Charlie gives a brilliantly inventive solo performance as a drunk returning home and finding things too much to cope with; *A Night Out* (1915); *The Rounders* (1914); *A Night in the Show* (1915).
Spectrum   VHS, Beta   B

Billy Liar

The Captain's Table

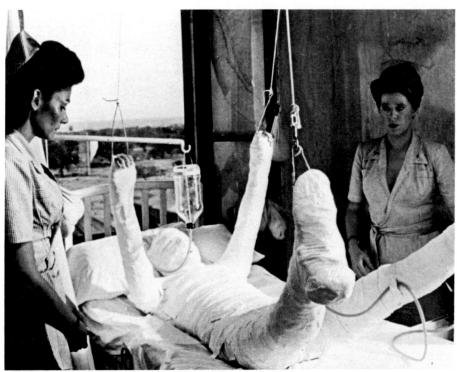

Catch 22

The Circus*

US 1928 72m ☐ bw

Charles Chaplin, Merna Kennedy, Allan Garcia, Harry Crocker, Henry Bergman
D, *Charles Chaplin*

Made at a difficult period in Chaplin's personal life and the least successful of his silent features, although enjoyable and the ending is memorable. With the 1919 two-reeler *A Day's Pleasure*.
Spectrum VHS, Beta A

City Lights***

US 1931 88m ☐ bw

Charles Chaplin, Virginia Cherrill, Harry Myers, Henry Bergman, Hank Mann
D, *Charles Chaplin*

If *The Gold Rush* is the best known Chaplin film, *City Lights* has been no less popular, triumphantly vindicating Chaplin's decision to persist with a silent comedy after the cinema had changed over to sound. But the decision was a difficult one and at one stage Chaplin stopped production while he weighed his dilemma. This partly explains why the film was nearly three years in the making but there were other delays, mainly due to Chaplin's way of working. On *City Lights* he abandoned the set for days on end as he waited for invention to return; he planned and shot elaborate sequences, only to discard them from the finished film; and halfway through he became so dissatisfied with his leading lady, Virginia Cherrill, that he sacked her; but he changed his mind and she returned, at a higher salary. Happily, these troubles are not evident on the screen and *City Lights* can claim to be the peak of Chaplin's artistic achievement. It is the story of a blind flower seller who befriends the little tramp, thinking him to be a millionaire; and of his efforts to help her regain her sight, while realising that if she does so the pretence must end. It is probably the most smoothly constructed of Chaplin's films, with the pathos and the comic business always at the service of the plot.
Spectrum VHS, Beta, CED A

College**
US 1927 65m □ bw/tint

*Buster Keaton, Ann Cornwall, Harold
Goodwin, Snitz Edwards
D, James W Horne and Buster Keaton*

If not quite in the highest rung of Keaton
films, it still offers a rich selection of his
exquisitely structured and impeccably
executed gags as he tries unsuccessfully to
prove himself as a student athlete – and get
the girl.
Spectrum   VHS, Beta   B

Cul-de-Sac*
GB 1966 105m ● bw

*Lionel Stander, Donald Pleasence, Jack
McGowran, Françoise Dorléac
D, Roman Polanski*

A distinctly macabre comedy in which two
gangsters on the run find themselves
sharing an island castle off Northumber-
land with a businessman-cum-artist and
his delectable young wife. An acquired
taste.
Videomedia   VHS, Beta   B

Dead Men Don't Wear Plaid*
US 1981 88m ● colour

*Steve Martin, Rachel Ward, Carl Reiner,
Reni Santoni
D, Carl Reiner*

Private eye spoof based on the ingenious
idea of helping the story along by inter-
cutting clips of classic 1940s thrillers
featuring such stars as Bogart, Stanwyck
and Alan Ladd. A good try.
Rank   VHS, Beta   A

Le Dernier Milliardaire*
France 1934 90m v bw

*Max Dearly, Renée Saint-Cyr, Marthe
Mellot, Raymond Cordy
D, René Clair*

Slight but engaging comic fantasy from a
director rather below top form. The queen
of a bankrupt kingdom tries to get her
daughter married off to a millionaire but
her plan is upset when the girl elopes with
a bandleader.
Thorn EMI   VHS, Beta   B

Diner*
US 1982 106m ● colour

*Steve Guttenberg, Daniel Stern, Mickey
Rourke, Kevin Bacon, Timothy Daly
D, Barry Levinson*

Witty and immaculately played piece in
which a group of lads just out of college
gather in a Baltimore diner and review their
lives, notably their success and otherwise
with girls. Neatly evocative of its 1959
period.
MGM/UA   VHS, Beta, CED   A

Doctor in the House*
GB 1954 92m □ colour

Dirk Bogarde, Kenneth More
D, Ralph Thomas

Original cinema version of Richard Gordon's medical student antics. It still comes up fresh and James Robertson Justice's irascible surgeon is richly comic. Also available: *Doctor at Large* (1957), *Doctor at Sea* (1955), *Doctor in Distress* (1963), *Doctor in Love* (1960), *Doctor in Clover* (1966), *Doctor in Trouble* (1970).
Rank VHS, Beta A

Duck Soup***
US 1933 70m □ bw

Groucho, Harpo, Chico and Zeppo Marx,
Margaret Dumont, Louis Calhern
D, Leo McCarey

For many fans, the best Marx Brothers romp of all: Groucho as Rufus T Firefly, ruler of Freedonia, outwitting enemy spies Chico and Harpo and winning the less than dainty hand of Margaret Dumont's statuesque Mrs Teasdale.
CIC VHS, Beta B

Eating Raoul*
US 1982 87m ● colour

Paul Bartel, Mary Woronow, Robert
Beltran, Susan Saiger, Ed Begley jr
D, Paul Bartel

Original and clever black comedy. A respectable middle class couple yield to greed and embark upon a series of murders to pay for their dream of a country restaurant; but a third party finds out.
Virgin VHS, Beta A

Entertaining Mr Sloane
GB 1969 90m ● colour

Beryl Reid, Harry Andrews, Peter
McEnery, Alan Webb
D, Douglas Hickox

Joe Orton's outrageous black comedy transferred from stage to screen with much of its flavour intact and rich performances by Reid and Andrews as a middle aged brother and sister competing for the favours of their young lodger.
Thorn EMI VHS, Beta A

Everything You Always Wanted To Know About Sex*

US 1972 88m ● colour

Woody Allen, Lynn Redgrave, John Carradine, Louise Lasser, Gene Wilder
D, Woody Allen

Seven sketches, of variable quality but containing enough wit and inspiration to be worth watching. Based on a sex manual, it was originally considered daring.
Warner VHS, Beta A

The Extra Girl*

US 1923 65m ☐ bw/tint

Mabel Normand, Max Davidson, Ralph Graves, George Nicholls
D, Mack Sennett

Charming silent comedy in which Sennett largely resists his usual slapstick and presents an excellent vehicle for his favourite comedienne, Mabel Normand; she plays a small-town beauty queen who tries her luck in Hollywood.
Spectrum VHS, Beta B

The Fast Lady

GB 1962 94m v colour

Stanley Baxter, James Robertson Justice, Leslie Phillips, Kathleen Harrison, Julie Christie
D, Ken Annakin

The Fast Lady is a vintage Bentley bought by a reluctant motorist to impress his girl friend – but he has not passed his driving test. Amiable concoction of car jokes and an early film part for Julie Christie.
Rank VHS, Beta A

First Monday in October

US 1981 99m v colour

Jill Clayburgh, Walter Matthau, Bernard Hughes, Jan Sterling, James Stephens
D, Ronald Neame

Clayburgh strikes a blow for feminism by becoming the first woman to be appointed to the US Supreme Court; Matthau plays the old adversary spoiling for fresh fights. Fairly enjoyable without generating the expected sparks.
CIC VHS, Beta A

Foul Play

US 1978 116m v colour

Goldie Hawn, Chevy Chase, Burgess Meredith, Rachel Roberts, Eugene Roche
D, Colin Higgins

Mixture of comedy and suspense, rather a long way after Hitchcock. Hawn as a dotty librarian getting wind of a plot to assassinate the Pope in San Francisco; Chase as the detective she has to convince.
CIC VHS, Beta A

Freebie and the Bean

US 1974 111m ● colour

Alan Arkin, James Caan, Loretta Swit, Jack Kruschen
D, Richard Rush

Frantic comedy-thriller about two hapless policemen – nicely played by Caan and Arkin – trying to corner a mobster. Noisy, violent and destructive, with a car smashing orgy involving more than 20 vehicles.
Warner VHS, Beta A

The Frisco Kid

US 1979 119m v colour

Gene Wilder, Harrison Ford, Ramon
Bieri, Leo Fuchs, Penny Peyser
D, Robert Aldrich

Wilder is a simple Polish rabbi crossing the American West on his way to San Francisco in the 1850s; Ford is an outlaw who becomes his friend and protector. A promising idea, let down by sentimentality and a rambling narrative.
Warner VHS, Beta A

The Gang That Couldn't Shoot Straight

US 1971 93m v colour

Jerry Orbach, Leigh Taylor-Young, Jo
Van Fleet, Lionel Stander
D, James Goldstone

Frantic black comedy about a Mafia war in New York, involving a cycle race, an unwanted lion and old pros Stander and Van Fleet hamming for all they are worth. More pace than wit.
MGM/UA VHS, Beta A

The General***

US 1926 76m □ bw

Buster Keaton, Marion Mack, Glen
Cavander, Jim Farley
D, Buster Keaton/Clyde Bruckman

A locomotive chase, based on an incident during the American Civil War, provides some of Keaton's finest comedy in a perfectly paced, strikingly photographed film. An undisputed cinema masterpiece.
Spectrum VHS, Beta B

Genevieve***

GB 1953 86m □ colour

Kenneth More, Kay Kendall, John
Gregson, Dinah Sheridan
D, Henry Cornelius

Rivalry between veteran car enthusiasts in the London to Brighton run is the framework for one of the happiest and most enduringly popular British films. Cast in cracking form and a catchy harmonica score by Larry Adler.
Rank VHS, Beta A

The Ghost Train

GB 1941 83m v bw

Arthur Askey, Richard Murdoch,
Kathleen Harrison, Morland Graham
D, Walter Forde

Third screen version of Arnold Ridley's comedy-thriller about a group of passengers stranded at a haunted station. The leading character was split to accommodate the famous radio partnership of Askey and Murdoch.
Rank VHS, Beta A

Foul Play

Genevieve

The Gold Rush

The Gold Rush***
US   1925   92m   ☐   bw

*Charles Chaplin, Georgia Hale, Mack
Swain, Tom Murray
D, Charles Chaplin*

Charlie, prospecting for gold in the Klon-
dike, has to cook and eat a shoe to stave off
hunger; and there are equally brilliant
sequences in one of Chaplin's finest works.
With the 1922 two-reeler, *Pay Day.*
Spectrum   VHS, Beta   A

The Goodbye Girl
US   1977   110m   v   colour

*Richard Dreyfuss, Marsha Mason, Quinn
Cummings, Paul Benedict
D, Herbert Ross*

Romantic comedy from Neil Simon about a
Broadway dancer abandoned by her husband and then her lover. She is reluctant to
start a new relationship when an aspiring
actor acquires the lease of her flat and
moves in.
Warner   VHS, Beta   A

The Graduate**
US   1967   105m   ●   colour

*Dustin Hoffman, Anne Bancroft,
Katharine Ross, Murray Hamilton
D, Mike Nichols*

Benjamin (Hoffman) returns home from
college to be seduced by the wife of his
father's partner (Bancroft). Despite abrupt
shifts of mood this is one of the sharpest
American comedies of the 1960s, brilliantly written and acted.
Warner   VHS, Beta   A

The Great Dictator***
US   1940   128m   v   bw

*Charles Chaplin, Paulette Goddard, Jack
Oakie, Reginald Gardiner
D, Charles Chaplin*

Satire on Hitler, with Chaplin playing both
the dictator and a persecuted Jewish
barber; some Chaplin admirers find it too
solemn, but many of the shots strike home.
Oakie makes a splendid Mussolini.
Spectrum   VHS, Beta, CED   A

The Great Race*
US   1965   147m   ☐   colour

*Jack Lemmon, Tony Curtis, Peter Falk,
Natalie Wood, George Macready
D, Blake Edwards*

Big-budget spectacular based on the first
New York to Paris car race in 1908, with
Lemmon and Curtis as rivals determined to
stop the other winning. Enjoyable set
pieces, but tries too hard at times.
Warner   VHS, Beta   A

Gregory's Girl**
GB   1980   91m   v   colour

*Gordon John Sinclair, Dee Hepburn,
Jake D'Arcy, Claire Grogan
D, Bill Forsyth*

A daydreaming schoolboy's attempt to
come to terms with the opposite sex in a
Scottish new town. Writer/director Forsyth's engaging comedy is well-observed
and perceptively true to life.
Video Movies   VHS, Beta   A

The Gumball Rally
US 1976 106m v colour

Michael Sarrazin, Norman Burton, Gary Busey, John Durren, Susan Flannery
D, Chuck Bail

An illegal coast-to-coast car rally organised by a bored New York executive for himself and his pals is the excuse for a predictable concoction of crashes and smashes, masterminded by a director who is also a stunt man.
Warner   VHS, Beta   A

Harry and Walter Go to New York
US 1976 107m □ colour

James Caan, Elliott Gould, Michael Caine, Diane Keaton, Charles Durning
D, Mark Rydell

Caan and Gould as failed vaudeville artists blasting themselves out of jail and trying to beat a rival to robbing a bank. Set in the 1890s, with nice period detail, but lacks the timing to make the most of the material.
RCA/Columbia   VHS, Beta   A

Heaven Can Wait
US 1978 100m v colour

Warren Beatty, Julie Christie, James Mason, Jack Warden, Charles Grodin
D, Warren Beatty/Buck Henry

Beatty is a footballer in heaven after a car accident. He returns to Earth to find that his body has been cremated and has to assume another identity. Tricky subject, which the film never quite masters.
CIC   VHS, Beta   A

Heavens Above!
GB 1963 113m v bw

Peter Sellers, Isabel Jeans, Cecil Parker, Eric Sykes, Brock Peters
D, John Boulting

Boulting Brothers' comedy of muddled intentions in which a north-country parson with naive ideas about charity is appointed to the wrong living and unwittingly causes mayhem. Sellers is much better than the film.
Thorn EMI   VHS, Beta   C

High Anxiety
US 1977 94m v colour

Mel Brooks, Madeline Kahn, Cloris Leachman, Harvey Korman, Ron Carey
D, Mel Brooks

Dedicated to the "Master of Suspense, Alfred Hitchcock" and consisting of a series of parodies of famous Hitchcock sequences, including the shower bath murder in *Psycho* (qv). Clever in part; patchy overall.
CBS/Fox   VHS, Beta, V2000, Laser   A

History of the World Part I
US   1981   92m   v   colour

Mel Brooks, Dom De Luise, Madeline Kahn, Cloris Leachman, Harvey Korman
D, *Mel Brooks*

Historical events from the Stone Age to the French Revolution given the free-wheeling, lavatorial Brooks treatment; either you love him or loathe him and this film provides ammunition for both sides.
CBS/Fox   VHS, Beta, V2000, Laser   A

Honky Tonk Freeway
US   1981   102m   v   colour

William Devane, Beau Bridges, Teri Garr, Beverly D'Angelo, Hume Cronyn
D, *John Schlesinger*

Madcap satire lost in chaos and based on the American obsession with the auto-mobile. An assortment of travellers are diverted to a small Florida resort whose mayor is desperate not to lose the tourist trade.
Thorn EMI   VHS, Beta   A

Hopscotch
US   1980   100m   v   colour

Walter Matthau, Glenda Jackson, Ned Beatty, Sam Waterston, Herbert Lom
D, *Ronald Neame*

Matthau as a CIA agent kicking over the traces by revealing all in his memoirs and Jackson his ally and mistress. Amiable spy spoof that could have been more effective with a sharper script and crisper handling.
Thorn EMI   VHS, Beta   A

How to Marry a Millionaire*
US   1953   93m   v   colour

Lauren Bacall, Marilyn Monroe, Betty Grable, William Powell
D, *Jean Negulesco*

Resurrection of the familiar gold-diggers plot as Bacall, Monroe and Gable rent a New York apartment in search of wealthy husbands. Fresh performances, particu-larly from Powell, help to make up for slack direction.
CBS/Fox   VHS, Beta   A

The Howling*
US   1980   90m   ●   colour

Dee Wallace, Patrick Macnee, Dennis Dugan, Christopher Stone
D, *Joe Dante*

Lively and inventive send-up of the horror film genre. Lots of in-jokes for buffs with the knowledge to appreciate them, but enough genuinely scary moments to satisfy those with a taste for the real thing.
Embassy   VHS, Beta   A

Hue and Cry**
GB 1946 78m v bw

Alastair Sim, Jack Warner, Harry Fowler,
Valerie White, Frederick Piper
D, Charles Crichton

A bunch of boys from the East End of London track down a criminal gang in this first Ealing comedy. Engaging and original, with plenty of pace and striking use of dockland locations.
Thorn EMI   VHS, Beta   C

An Ideal Husband
GB 1947 91m v colour

Paulette Goddard, Hugh Williams,
Michael Wilding, Diana Wynyard
D, Alexander Korda

Oscar Wilde's play about a British diplomat whose career and private happiness are threatened by the reappearance of an old flame. Visually striking, with strong decor and elegant costumes, but dramatically stilted.
Spectrum   VHS, Beta   A

I'm All Right Jack*
GB 1959 101m v bw

Peter Sellers, Ian Carmichael, Irene
Handl, Terry-Thomas, Liz Fraser
D, John Boulting

This swipe at management and trade unions in British industry contains a magnificent portrait by Sellers, observed down to the last detail, of a Communist shop steward; otherwise it is somewhat crude farce.
Thorn EMI   VHS, Beta   C

The Importance of Being Earnest**
GB 1952 95m □ colour

Michael Redgrave, Michael Denison,
Edith Evans, Margaret Rutherford
D, Anthony Asquith

Stylish and theatrical rendering of Oscar Wilde's wittiest comedy. A cluster of good performances includes the definitive Lady Bracknell of her generation by Dame Edith Evans.
Rank   VHS, Beta   A

It's a Mad, Mad, Mad, Mad World*
US 1963 162m □ colour

Spencer Tracy, Jimmy Durante, Milton
Berle, Sid Caesar, Ethel Merman
D, Stanley Kramer

A frenetic search for buried loot, with car chases and other elaborate set pieces, in what set out to be the longest and most star-studded comedy ever. Good moments are matched by excesses.
Warner   VHS, Beta   A

Jour de Fête***
France 1947 80m □ bw

Jacques Tati, Guy Decombe, Paul
Frankeur, Santa Relli
D, Jacques Tati

Inspired clowning by Tati as a village postman striving for efficiency but hampered by an old bicycle and his own clumsiness. The gags are combined with an affectionate look at French rural life.
Videomedia   VHS, Beta, V2000   B

The Kid***
US 1921 90m □ bw

Charles Chaplin, Jackie Coogan, Edna Purviance, Carl Miller, Tom Wilson
D, Charles Chaplin

Chaplin drew on his harsh upbringing in the London slums for this poignant tale about a tramp who befriends an abandoned child; he also coaxed an unforgettable performance from the five-year-old Coogan. With *The Idle Class* (1921).
Spectrum VHS, Beta A

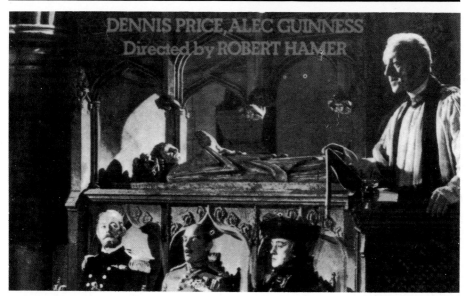

Kind Hearts and Coronets***
GB 1949 102m □ bw

Dennis Price, Alec Guinness, Valerie Hobson, Joan Greenwood, Miles Malleson
D, Robert Hamer

Most Ealing comedies were comfortable, affectionate celebrations of the British character with no intention to disturb or shock. *Kind Hearts and Coronets* stands completely outside this tradition; it is a black comedy of murder, ironic and subversive, which still has virtually no parallel in the British cinema. It was suggested to Robert Hamer, the director, by an Edwardian novel, *Israel Rank,* written by a follower of Oscar Wilde, Roy Horniman. The hero (or rather anti-hero) Louis Mazzini is descended from the Duke of Chalfont but is reduced to working in a suburban draper's shop because the family disowned his mother for eloping with an Italian singer. Louis determines to recover the family title but to do so he must dispose of the relatives who stand in his way. The film is most remembered for the versatility of Alec Guinness, who plays all eight victims, but Price is equally good as the suave, calculating Louis. It is also notable for the wit of its dialogue and there are many visual felicities to dispute the view that, for all its brilliance, the film is fundamentally uncinematic. This will be a lasting monument to Hamer who never again produced a film of remotely the same quality and died at the early age of 52.
Thorn EMI VHS, Beta C

The Ladykillers***
GB   1955   87m   v   colour

*Alec Guinness, Katie Johnson, Cecil
Parker, Herbert Lom, Danny Green
D, Alexander Mackendrick*

Sweet old lady (marvellous Katie Johnson)
unwittingly harbours a gang of villains
passing themselves off as a string quartet
as they plan a wages snatch at King's Cross
Station. Ealing comedy bristling with sharp-
edged wit.
Thorn EMI   VHS, Beta   C

The Lavender Hill Mob***
GB   1951   78m   □   bw

*Alec Guinness, Stanley Holloway, Sidney
James, Alfie Bass, Marjorie Fielding
D, Charles Crichton*

Guinness, a timid bank clerk, masterminds
a daring bullion robbery from his own bank.
A very likeable and immaculately plotted
Ealing comedy that has the great virtue of
not going on a second too long.
Thorn EMI   VHS, Beta   C

Let George Do It*
GB   1940   81m   □   bw

*George Formby, Phyllis Calvert, Garry
Marsh, Romney Brent, Bernard Lee
D, Marcel Varnel*

The simple charm of the toothy, ukelele-
strumming Formby helped to make him for
five years running the most popular star of
the British cinema. Here he is mistaken by
the Nazis for a spy in Norway.
Thorn EMI   VHS, Beta   C

Limelight***
US   1952   145m   v   bw

*Charles Chaplin, Claire Bloom, Sydney
Chaplin, Nigel Bruce, Buster Keaton
D; Charles Chaplin*

Touching piece about a fading music hall
comedian who saves a young ballerina
from suicide and by encouraging her career
puts new meaning into his own. The
Chaplin–Keaton routine is a gem of comic
timing.
Spectrum   VHS, Beta   A

Little Miss Marker
US   1980   99m   □   colour

*Walter Matthau, Julie Andrews, Tony
Curtis, Bob Newhart, Sara Stimson
D, Walter Bernstein*

Matthau in a tailor-made part as a cynical
New York bookmaker melted by the cute
little girl he is left to look after by a client.
Efficient fourth screen version of a 1930s
Runyon short story.
CIC   VHS, Beta   A

Local Hero**
GB   1983   107m   v   colour

*Burt Lancaster, Denis Lawson, Fulton
Mackay, Jenny Seagrove, Peter Reigert
D, Bill Forsyth*

Old and new cultures collide when an
American oil company plans to build a
refinery in a Scottish village and finds itself
up against the canny locals. Full of subtle,
unforced humour arising naturally from
character and situation.
Thorn EMI   VHS, Beta   A

Loot*
GB   1970   96m   ●   colour

*Richard Attenborough, Lee Remick,
Hywel Bennett, Milo O'Shea
D, Silvio Narizzano*

Adaptation by Ray Galton and Alan Simp-
son of Joe Orton's black farce about a
young crook who hides the proceeds of a
bank robbery in his mother's coffin. Stage
play developed boldly and at a furious pace.
Thorn EMI   VHS, Beta   A

Lost and Found
US   1979   102m   ●   colour

*Glenda Jackson, George Segal, Maureen
Stapleton, Paul Sorvino, Hollis McLaren
D, Melvin Frank*

Less-than-happy revival of the Jackson–
Segal partnership after its great popular
success in *A Touch of Class* (qv). An English
divorcée and an American professor meet
on a skiing holiday, marry and find themselves at odds.
RCA/Columbia   VHS, Beta   A

Love and Death*
US   1975   83m   v   colour

*Woody Allen, Diane Keaton, Georges
Adel, Frank Adu
D, Woody Allen*

Allen's idiosyncratic lampoon of *War and
Peace*. He plays a Russian condemned to
death for shooting Napoleon; but, as in
other Allen films, the often splendid gags
are more memorable than the whole.
Warner   VHS, Beta   A

Love at First Bite*
US   1979   96m   v   colour

*George Hamilton, Susan Saint James,
Richard Benjamin, Dick Shawn
D, Stan Dragoti*

When his Transylvanian castle is taken
over by the government as a gymnasium,
Count Dracula moves to New York and falls
in love with a model. Highly successful
combination of mock horror and old-
fashioned romantic comedy.
Guild   VHS, Beta   A

Lovers and Other Strangers*
US   1969   91m   ●   colour

*Bonnie Bedelia, Gig Young, Michael
Brandon, Beatrice Arthur, Richard
Castellano
D, Cy Howard*

A couple who have lived together for 18
months decide to get married; but their
relatives at the wedding find their own sex
problems less easy to resolve. A witty script
and several excellent performances.
Rank   VHS, Beta   A

Loving Couples
US   1980   98m   ●   colour

*Shirley Maclaine, James Coburn, Susan
Sarandon, Stephen Collins
D, Jack Smight*

An attempt at a permissive comedy with
Maclaine and Coburn as doctors who seek
consolation for their unsuccessful marriage by having affairs but end up rejected
by their new partners. Could have been
better.
VTC   VHS, Beta   A

Lucky Jim*
GB   1957   91m   v   bw

*Ian Carmichael, Hugh Griffith, Terry-
Thomas, Sharon Acker, Jean Anderson
D, John Boulting*

Satire turns into farce as the Boulting
brothers give their typical treatment to the
novel by Kingsley Amis about a hapless
young lecturer at a provincial university.
The story is still funny despite a loss of
subtlety.
Thorn EMI   VHS, Beta   C

Hue and Cry

Jour de Fête

Love at First Bite

The Maggie*
GB 1954 90m ☐ bw

*Paul Douglas, Alex Mackenzie, James
Copeland, Abe Barker, Dorothy Alison
D, Alexander Mackendrick*

A lesser Ealing comedy, though nicely
played and with good location work, about
the humiliation of an American business
tycoon who foolishly entrusts his furniture
to the skipper of a dilapidated cargo boat on
the Clyde.
Thorn EMI   VHS, Beta   C

The Magic Christian
GB 1969 95m v colour

*Peter Sellers, Ringo Starr, Richard
Attenborough, Laurence Harvey
D, Joseph McGrath*

Sellers and Starr as an eccentric millionaire and his adopted son whose delight is to
play tricks on the rich and favoured.
Disjointed adaptation of Terry Southern's
novel, providing cameos for a host of
British comedy actors.
Video Form   VHS, Beta   A

The Man in the White Suit***
GB 1951 82m v bw

*Alec Guinness, Joan Greenwood, Cecil
Parker, Vida Hope, Ernest Thesiger
D, Alexander Mackendrick*

Ealing comedy at its most astringent.
Guinness is the humble employee of a
textile factory who discovers an everlasting
fabric, but both management and unions
see it as a direct threat to their livelihoods.
Thorn EMI   VHS, Beta   B

M*A*S*H***
US 1970 111m ● colour

*Donald Sutherland, Elliott Gould, Tom
Skerritt, Sally Kellerman, Robert Duvall
D, Robert Altman*

Outrageously funny black comedy about
American army surgeons tending the
wounded behind the lines during the
Korean War. They seek relief from the
grimness of their task by chasing women
and thumbing their noses at authority.
CBS/Fox   VHS, Beta, V2000, Laser   A

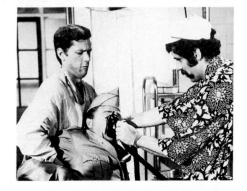

A Midsummer Night's Sex Comedy
US 1982 85m ● colour

Woody Allen, Mia Farrow, José Ferrer
D, Woody Allen

At a weekend house party at a stock-broker's country retreat six couples review and try to resolve their sex problems. Though set at the turn of the century, this comedy of neurosis has a clear contemporary ring.
Warner VHS, Beta A

The Missionary*
GB 1981 82m v colour

Michael Palin, Maggie Smith, Trevor Howard, Denholm Elliott, Graham Crowden
D, Richard Loncraine

Palin (who also wrote the script) is an Edwardian clergyman who sets up a mission for fallen women in the East End of London. He provides more than just shelter – and a few laughs.
Thorn EMI VHS, Beta, CED A

Mr Blandings Builds His Dream House**
US 1948 90m v bw

Cary Grant, Myrna Loy, Melvyn Douglas
D, H C Potter

The tribulations of a New York advertising man and his family pursuing their dream of living in the country by buying a ramshackle house in Connecticut. With *Mexican Spitfire at Sea* (70m), a 1941 second feature comedy.
Kingston VHS, Beta A

Mr Roberts*
US 1955 115m ☐ colour

Henry Fonda, James Cagney, William Powell, Jack Lemmon, Betsy Palmer
D, John Ford and Mervyn Le Roy

Rather protracted version of a long-running Broadway play about a man (Fonda) who wants to see action in the Second World War but is stuck on a cargo ship; Cagney is the eccentric captain, Powell the doctor.
Warner VHS, Beta A

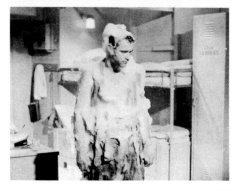

Modern Times***

US 1936 89m □ bw

Charles Chaplin, Paulette Goddard,
Henry Bergman, Chester Conklin
D, Charles Chaplin

Charlie as a factory worker caught up in the inhumanity of the assembly line and eventually unhinged by it; appealing performance by Goddard as the gamine who befriends him and offers a ray of hope in an unfriendly world.
Spectrum VHS, Beta A

Mon Oncle**

France 1958 105m □ colour

Jacques Tati, Jean-Pierre Zola, Alain
Becourt, Lucien Fregis
D, Jacques Tati

M Hulot (see below) reappears as the uncle of a seven-year-old boy, trying without much success, to make sense of modern gadgetry. Sublime moments in a gentle, amused look at the technological age, if a little slow.
Videomedia VHS, Beta, V2000 B

Monsieur Hulot's Holiday***

France 1953 91m □ bw

Jacques Tati, Nathalie Pascaud, Michèle
Rolla, Valentine Camax
D, Jacques Tati

The amiable, innocent, accident-prone M Hulot – with pipe, jaunty hat and stiff-legged walk – is one of the cinema's unforgettable comic creations. The setting here is a small seaside resort in Brittany.
Videomedia VHS, Beta, V2000 B

Monty Python and the Holy Grail*

GB 1974 90m v colour

Graham Chapman, John Cleese, Terry
Gilliam, Eric Idle, Terry Jones, Michael
Palin
D, Terry Gilliam and Terry Jones

The Monty Python television team in a medieval romp based loosely on King Arthur and the Round Table. Essentially a series of sketches, which strike a chord.
Brent Walker VHS, Beta A

Monty Python's Life of Brian*

GB 1979 89m v colour

Terry Jones, Graham Chapman, Michael Palin, John Cleese, Eric Idle
D, Terry Jones

Spirited spoof on the story of Jesus which, according to your point of view, is either a tasteless blasphemy or a delightfully irreverent tilt at religious pomposity, particularly in the Hollywood biblical epics.
Thorn EMI   VHS, Beta   A

Morgan – A Suitable Case For Treatment*

GB 1966 93m v bw

Vanessa Redgrave, David Warner, Robert Stephens, Irene Handl
D, Karel Reisz

Skilfully played by the lanky Warner, Morgan is an artist whose inability to cope with life leads him to retreat into fantasy. A comic treatment of madness which is often more disturbing than funny.
Thorn EMI   VHS, Beta   C

The Mouse That Roared*

GB 1959 80m v colour

Peter Sellers, Jean Seberg, David Kossoff, William Hartnell, Leo McKern
D, Jack Arnold

The bankrupt duchy of Grand Fenwick hopes to solve its economic problems by declaring war on the United States and receiving aid as a defeated power. The comic premise has dated but not the skill of Sellers, who plays three parts.
RCA/Columbia   VHS, Beta   A

Movie Movie*

US 1978 102m v c/bw

George C Scott, Trish van Devere, Red Buttons, Eli Wallach, Barbara Harris
D, Stanley Donen

Pastiche of a cinema double bill from the 1930s comprising *Dynamite Hands*, a black and white boxing melodrama, and *Baxter's Beauties of 1933*, a backstage musical. Cleverly done, but connoisseurs might prefer the real thing.
Precision   VHS, Beta   A

Murder By Death

US 1976 94m v colour

*Peter Falk, Alec Guinness, Peter Sellers,
Elsa Lanchester, David Niven
D, Robert Moore*

The world's greatest fictional detectives are invited to a weekend house party with $1m on offer to the sleuth who can solve a murder due to take place at midnight. Good performances, but the parody lacks substance.

RCA/Columbia VHS, Beta R

My Favourite Year*

US 1982 89m v colour

*Peter O'Toole, Mark Linn-Baker, Jessica Harper, Joseph Bologna
D, Richard Benjamin*

O'Toole as a hard-drinking, hell-raising actor (shades of Errol Flynn, not to mention O'Toole himself) trying to pull himself together for a television appearance. Nicely judged humour and excellent performances.

MGM/UA VHS, Beta, Laser, CED A

My Learned Friend**

GB 1943 71m v bw

*Will Hay, Claude Hulbert, Mervyn Johns,
Ernest Thesiger, Charles Victor
D, Will Hay and Basil Dearden*

Hay as a disreputable lawyer who finds himself on the death list of an ex-convict. Several memorable sequences in fast-moving black farce, which marked a much-loved comic's farewell to the cinema.

Thorn EMI VHS, Beta C

National Lampoon's Animal House

US 1978 106m v colour

*John Belushi, Tim Matheson, John
Vernon, Verna Bloom, Donald Sutherland
D, John Landis*

American campus comedy, set in the early 1960s, which was enormously successful in the United States but is likely to mean less to audiences elsewhere. *National Lampoon* is a monthly magazine.

CIC VHS, Beta A

Nickelodeon

US/GB 1976 117m v colour

*Ryan O'Neal, Burt Reynolds, Tatum
O'Neal, Brian Keith, Stella Stevens
D, Peter Bogdanovich*

For a man steeped in the lore of the cinema, Bogdanovich should have made more of this light-hearted look at pioneer movie-making in Hollywood. A promising subject comes out disappointingly flat.

Thorn EMI VHS, Beta A

A Night at the Opera***
US   1935   88m   v   bw

Groucho, Chico and Harpo Marx, Kitty Carlisle, Allan Jones, Margaret Dumont
D, Sam Wood

Everyone has their favourite Marx Brothers film, but this must be near the top of most lists. Groucho bids to save a failing opera house with the help of the wealthy Mrs Claypool; the invention seldom flags.
MGM/UA   VHS, Beta   A

Nine to Five
US   1980   110m   v   colour

Jane Fonda, Lily Tomlin, Dolly Parton, Dabney Coleman, Sterling Hayden
D, Colin Higgins

Three office girls plot to turn the tables on their chauvinistic boss. A promising idea, it starts crisply enough but is ultimately let down by a weak script. Tomlin is easily the pick of the trio.
CBS/Fox   VHS, Beta   A

1941*
US   1979   114m   v   colour

Dan Aykroyd, Ned Beatty, John Belushi, Lorraine Gary, Christopher Lee
D, Steven Spielberg

A Japanese invasion of California in the aftermath of Pearl Harbour provokes the population to hysteria. Said to be the most expensive comedy ever made, but proves that a big budget is no substitute for wit and style.
CIC   VHS, Beta   A

Ocean's Eleven
US   1960   127m   v   colour

Frank Sinatra, Peter Lawford, Sammy Davis Jnr, Dean Martin, Angie Dickinson
D, Lewis Milestone

Uneven comedy-thriller featuring the Sinatra "clan" as a gang planning to rob a casino in Las Vegas. The performers are clearly enjoying themselves but the audience may sometimes find it hard to share the jokes.
Warner   VHS, Beta   A

Oh, God!*
US   1977   99m   v   colour

George Burns, John Denver, Ralph Bellamy, Donald Pleasence, Teri Garr
D, Carl Reiner

Assistant supermarket manager answers an invitation to meet God and encounters an old man in baseball cap and sneakers, Burns, and the veteran comedian delightfully makes the most of a trifle.
Warner   VHS, Beta   A
also available: *Oh, God! Book II* (Warner)

On the Beat/Trouble in Store
GB   1962/1953   188m   □   bw

Norman Wisdom, Jennifer Jayne, Raymond Huntley, Margaret Rutherford
D, Robert Asher and John Paddy Carstairs

Wisdom's Chaplinesque "little man" in two of the films which helped to make him one of Britain's biggest box-office draws. In the first he is a car park attendant, in the second an assistant in a store.
Rank   VHS, Beta   A

Only Two Can Play**
GB   1962   101m   ●   bw

*Peter Sellers, Mai Zetterling, Virginia
Maskell, Richard Attenborough
D, Sidney Gilliat*

Sellers in one of his best parts, as a put-upon Welsh librarian who tries to escape from his drab family surroundings by having an affair with the glamorous wife of a local councillor. Based on Kingsley Amis's novel, *That Uncertain Feeling*.
Thorn EMI   VHS, Beta   B

The Owl and the Pussycat*
US   1970   97m   ●   colour

*Barbra Streisand, George Segal, Robert
Klein, Allen Garfield, Roz Kelly
D, Herbert Ross*

Two-hander between Segal's bookshop assistant aspiring to be a writer and Streisand's self-styled model and actress. A standard sex comedy brought into the permissive age: vulgar, superbly acted, consistently funny.
RCA/Columbia   VHS, Beta   R

The Pink Panther*
US   1963   113m   □   colour

*Peter Sellers, David Niven, Capucine
D, Blake Edwards*

Sellers as the bumbling Inspector Clouseau on the trail of an international jewel thief in the first, and best, of a series.
Warner   VHS, Beta   A
also: *The Return of the Pink Panther* (1974, Precision), *The Pink Panther Strikes Again* (1976, Warner) and *The Revenge of the Pink Panther* (1978, Warner).

Play It Again, Sam*
US   1972   82m   v   colour

*Woody Allen, Diane Keaton, Jerry Lacy,
Susan Anspach
D, Herbert Ross*

Allen stars in an adaptation of his own play about a neurotic film critic who calls up the spirit of Humphrey Bogart to help his love life when his wife walks out. A good comic idea, particularly for film buffs.
CIC   VHS, Beta   B

Playtime*
France   1968   115m   v   colour

*Jacques Tati, Barbara Dennek,
Jacqueline Lecomte, Henri Piccoli
D, Jacques Tati*

Tati's M Hulot gets caught up with a group of American tourists in Paris and shares their bewilderment at the sameness of modern buildings. An amusing attack on the concrete-and-glass society, though the jokes are rather slow.
Videomedia   VHS, Beta   B

A Night at the Opera

Play it Again, Sam

The Prince and the Showgirl
GB 1957 117m v colour

Laurence Olivier, Marilyn Monroe, Sybil Thorndike, Richard Wattis
D, Laurence Olivier

Handsomely mounted but ponderously handled version of Terence Rattigan's romantic comedy of a suave European prince and a simple chorus girl who meet in London during the 1911 Coronation. Two fascinating star performances.
Warner VHS, Beta A

The Prisoner of Second Avenue*
US 1975 101m v colour

Jack Lemmon, Anne Bancroft, Gene Saks, Elizabeth Wilson
D, Melvin Frank

Lemmon and Bancroft make the most of some typically witty dialogue in an untypical Neil Simon comedy. An advertising man loses his job and has a nervous breakdown.
Warner VHS, Beta A

Private Benjamin
US 1980 108m v colour

Goldie Hawn, Eileen Brennan, Armand Assante, Robert Webber
D, Howard Zieff

Shocked by the death of her second husband on their wedding night, rich young widow Hawn seeks a new life by joining the army and comes up against a succession of male chauvinists. Muddled would-be feminist piece of variable quality.
Warner VHS, Beta A

Privates on Parade*
GB 1982 107m v colour

John Cleese, Denis Quilley, Michael Elphick, Nicola Pagett, Joe Melia
D, Michael Blakemore

Faithful screen treatment of Peter Nichols' semi-autobiographical stage hit about blackmail and intrigue among British army conscripts running an entertainment group in Singapore in 1948. Exuberant and bawdy with a nice sense of period.
Thorn EMI VHS, Beta, CED A

The Producers*
US 1967 84m v colour

Zero Mostel, Gene Wilder, Kenneth Mars, Estelle Winwood, Renee Taylor
D, Mel Brooks

Mostel as the devious Broadway producer who puts on a show called *Springtime for Hitler* which is so bad that it cannot possibly succeed – and sees it become a smash hit. Lovely satire that made Brooks into a cult.
Embassy VHS, Beta A

Puberty Blues

Australia 1981 85m ● colour

Nell Schofield, Jad Capelja, Geoff Rhoe, Tony Hughes, Sandy Paul
D, Bruce Beresford

Two teenage girls lose their virginity amid the Australian beach and surfboard culture. Comedy of sexual awakening that tries unconvincingly to make a general statement about the morals of modern youth.
Thorn EMI   VHS, Beta   A

Road to Utopia*

US 1945 86m v bw

Bob Hope, Bing Crosby, Dorothy Lamour, Douglass Dumbrille, Hillary Brooke
D, Hal Walker

Hope and Crosby as vaudevillians joining the California gold rush. The fourth of the "Road" series, with a good helping of gags, Lamour providing the usual romantic interest and Robert Benchley supplying a commentary.
CIC   VHS, Beta   B

Royal Flash

GB 1975 97m v colour

Malcolm McDowell, Alan Bates, Florinda Bolkan, Britt Ekland, Oliver Reed
D, Richard Lester

The odious Flashman, recreated by George MacDonald Fraser, bulldozes his boastful way round Europe. Handsome locations and elegant Victorian costumes are stunning, but not all the jokes are up to the same standard.
CBS/Fox   VHS, Beta, V2000, Laser   A

Sally of the Sawdust

US 1925 86m □ bw/tint

W C Fields, Carol Dempster, Alfred Lunt, Effie Shannon, Erville Anderson
D, D W Griffith

Fields made his film debut in this silent screen version of *Poppy*. He repeats his successful stage part as a devious but kind hearted circus juggler touring the country with his adopted daughter (Dempster).
Spectrum   VHS, Beta   B

Satan's Brew*

WG 1976 100m ● colour

Kurt Raab, Margit Carstensen, Helen Vita, Volker Spengler
D, Rainer Werner Fassbinder

Appropriately titled black comedy about a revolutionary poet with writer's block who believes himself to be the reincarnation of the great German Symbolist poet, Stefan George, and arranges his life accordingly.
Palace   VHS, Beta   A

Seems Like Old Times
US 1980 96m v colour

*Goldie Hawn, Chevy Chase, Charles
Grodin, Robert Guillaume, Harold Gould
D, Jay Sandrich*

A thirties-style crazy comedy, with Hawn
as a crusading Los Angeles lawyer giving
unwilling refuge to her ex-husband (Chase),
an accident-prone writer embroiled in a
bank robbery. Stream of jokes about cats,
dogs and servants.
RCA/Columbia VHS, Beta A

Semi-Tough*
US 1977 105m v colour

*Burt Reynolds, Kris Kristofferson, Jill
Clayburgh, Bert Convy, Robert Preston
D, Michael Ritchie*

Reynolds and Kristofferson as professional
footballers competing for the favours of the
much married daughter (Clayburgh) of the
team manager. Gently satirical piece, with
the American vogue for self-help therapy
the main target.
Warner VHS, Beta A

Silver Streak
US 1976 108m v colour

*Gene Wilder, Richard Pryor, Jill
Clayburgh, Patrick McGoohan
D, Arthur Hiller*

Romance and murder aboard the Los
Angeles to Chicago express. Undemanding
caper which curiously underuses two of its
stars – Clayburgh and Pryor – and leaves
Wilder, as a bemused publisher, to hold
things together.
CBS/Fox VHS, Beta A

Smokey and the Bandit*
US 1977 96m v colour

*Burt Reynolds, Jackie Gleason, Sally
Field, Jerry Reed, Mike Henry
D, Hal Needham*

Ace trucker Reynolds on a 900-mile round
trip with a consignment of beer, picking up
a damsel in distress and getting himself
chased by irate sheriff Gleason. Superior
car-smash frolic with good stunts. Also
available: *Smokey and the Bandit II.*
CIC VHS, Beta A

S.O.B.
US 1981 117m ● colour

*Julie Andrews, William Holden, Richard
Mulligan, Robert Vaughn, Robert Webber
D, Blake Edwards*

A Hollywood director with a disastrous flop
on his hands decides to remake it as a sex
movie. Film industry satire which often
descends to the vulgarity it is trying to
lampoon; it's the one in which Julie Andrews
goes briefly topless.
Guild VHS, Beta A

Stardust Memories*
US 1980 83m v bw

*Woody Allen, Charlotte Rampling,
Jessica Harper, Marie-Christine Barrault
D, Woody Allen*

Allen can hardly get more autobiographical
than this, playing a director of comedy films
beset by doubts about the value of his work
and trying to smooth the tangles of his
private life. A potent study of introspection.
Warner VHS, Beta A

Starting Over
US 1979 101m ● colour

*Burt Reynolds, Jill Clayburgh, Candice
Bergen, Charles Durning
D, Alan J. Pakula*

Neurotic comedy/romance in which
magazine writer Reynolds finds himself
having to decide whether to mend fences
with his ambitious ex-wife (Bergen) or start
afresh with a nursery school teacher
(Clayburgh). Psychiatry can't help him.
CIC VHS, Beta A

Stir Crazy
US 1980 107m ● colour

*Gene Wilder, Richard Pryor, George
Stanford Brown, JoBeth Williams
D, Sidney Poitier*

Wilder and Pryor giving full – perhaps too
much – vent to their comic talents as a
couple thrown out of work in New York who
head for California and land in prison.
Although engaging, it needed tighter script
and direction.
RCA/Columbia VHS, Beta, CED A

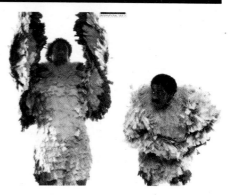

Stripes
US 1981 103m v colour
*Bill Murray, Harold Ramis, Warren
Oates, P J Soles, John Larroquette
D, Ivan Reitman*

Murray and co-author Ramis in sparkling
form as a couple of new recruits to a
shambling US army platoon licked into
shape by Oates's martinet sergeant. Director Reitman makes fresh use of well worn
material.
RCA/Columbia   VHS, Beta   R

The Sunshine Boys*
US 1975 99m v colour
*Walter Matthau, George Burns, Richard
Benjamin, Carol Arthur
D, Herbert Ross*

Adaptation of the Neil Simon Broadway hit
about two old vaudeville comedians being
persuaded out of retirement and resuming
their animosity. Burns (his first film for 36
years) is effortlessly brilliant, Matthau
uncomfortable and forced.
MGM/UA   VHS, Beta   A

Take the Money and Run
US 1969 83m v colour
*Woody Allen, Janet Margolin, Marcel
Hillaire, Jacqueline Hyde
D, Woody Allen*

Allen's first film as a director and making
an uncertain start: a comic version of the
old Hollywood standard about the innocent
forced by bad social conditions into a life of
crime. Some marvellous gags but they do
not cohere.
Rank   VHS, Beta   A

Taking Off**
US 1971 83m ● colour
*Lynn Carlin, Buck Henry, Linnea
Heacock, Georgia Engel, Tony Harvey
D, Milos Forman*

A beautifully observed and extremely witty
piece by Czech emigré Forman, about the
generation gap. Conventional American
parents go on the trail of their missing
teenage daughter and uncover a lifestyle
they do not comprehend.
CIC   VHS, Beta   B

"10"*

US 1979 118m ● colour

Dudley Moore, Bo Derek, Julie Andrews,
Robert Webber
D, Blake Edwards

Moore plays a middle-aged composer who indulges his sexual fantasies by rating girls from one to ten. Bo Derek is the dream girl he pursues to Mexico in a hilarious sequence; and Julie Andrews is the "steady".
Warner VHS, Beta A

That Sinking Feeling*

GB 1979 90m v colour

Tom Mannion, Eddie Burt, Richard
Demarco, Alex Mackenzie
D, Bill Forsyth

Young Glaswegians on the dole try to alleviate their boredom and frustration by stealing a consignment of stainless steel sinks from a warehouse. Fresh and inventive first feature by director of *Gregory's Girl* and *Local Hero* (qqv).
Palace VHS, Beta A

There's a Girl in my Soup

GB 1970 92m ● colour

Peter Sellers, Goldie Hawn, Tony Britton,
Nicky Henson, John Comer
D, Roy Boulting

Sellers as the amorous television gourmet of Terence Frisby's stage success, meeting his match when he tries out his seduction technique on a kooky young blonde. A mild comedy of humiliation which overstretches a thin plot.
RCA/Columbia VHS, Beta R

They're a Weird Mob

Australia/GB 1966 112m v colour

Walter Chiari, Clare Dunne, Chips
Rafferty, Alida Chelli, Ed Devereaux
D, Michael Powell

Likeable immigrant-makes-good story. Italian journalist, in Australia to work on his cousin's newspaper, discovers that the cousin has fled leaving large debts and has to work as a labourer to pay them off.
Rank VHS, Beta A

Those Magnificent Men in their Flying Machines*

GB 1965 127m □ colour

Sarah Miles, Stuart Whitman, James
Fox, Robert Morley, Eric Sykes
D, Ken Annakin

With so many well known faces in cameo parts it becomes a case of "spot the actor". The machines are the real stars of this amiable, extended knockabout comedy of the first London to Paris air race in 1910.
CBS/Fox VHS, Beta A

Time Bandits
GB 1981 105m v colour

John Cleese, Sean Connery, Ian Holm,
Ralph Richardson, David Warner
D, Terry Gilliam

Blackish comedy from the Monty Python
stable, with a strong current of violence.
The escapades of a schoolboy and a group
of dwarves as they make a tour through
history and encounter the likes of Robin
Hood, Napoleon and King Agamemnon.
Thorn EMI   VHS, Beta   A

The Titfield Thunderbolt*
GB 1952 80m □ colour

Stanley Holloway, George Relph, John
Gregson, Godfrey Tearle, Naunton
Wayne
D, Charles Crichton

Typical Ealing whimsy about villagers
fighting to save their branch railway,
threatened with closure, by running it
themselves. The colour photography is a
treat and so are the old puffers.
Thorn EMI   VHS, Beta   C

A Touch of Class*
GB 1973 106m v colour

Glenda Jackson, George Segal, Paul
Sorvino, Hildegarde Neil, Cec Linder
D, Melvin Frank

A chance encounter in London between a
dress designer and a married American
businessman leads to a whirlwind affair.
Predictably polished playing from Segal
and Jackson's spiky performance, in her
first comedy part, won her an Oscar.
CBS/Fox   VHS, Beta   B

The Toy
US 1982 97m v colour

Richard Pryor, Jackie Gleason, Ned
Beatty, Scott Schwartz, Teresa Ganzel
D, Richard Donner

Pryor as an out-of-work journalist doing
menial jobs for a big corporation who is
offered the post of "toy" to the boss's nine-
year-old son. Some inspired clowning by
the star.
RCA/Columbia   VHS, Beta   R

Two-Way Stretch*
GB 1960 84m v bw

Peter Sellers, Lionel Jeffries, Wilfrid Hyde White, Bernard Cribbins
D, Robert Day

Three convicts plan the perfect crime: to slip out of jail, rob a maharajah of his jewels, and get back inside before anyone notices. A cleverly plotted and enjoyable romp with a roster of British comedy talent at the top of its form.
Thorn EMI VHS, Beta C

Uptown Saturday Night
US 1974 100m v colour

Sidney Poitier, Bill Cosby, Harry Belafonte, Flip Wilson
D, Sidney Poitier

Frantic caper, bursting with energy but short on wit, in which a couple of pals track down a gang who have run off with a winning lottery ticket. The only novel feature is the all-black cast.
Warner VHS, Beta A

Victor/Victoria**
GB 1982 129m ● colour

Julie Andrews, James Garner, Robert Preston, Leslie-Anne Warren
D, Blake Edwards

Stylish high camp version of a German film comedy of the 1930s, with Andrews as a struggling singer who becomes an overnight success by masquerading as a drag artist. Preston superb as the ageing queen who puts her up to it.
MGM/UA VHS, Beta, V2000, Laser A

The Virgin Soldiers*
GB 1969 96m ● colour

Hywel Bennett, Nigel Patrick, Lynn Redgrave, Nigel Davenport
D, John Dexter

Cleaned-up but generally effective transcription of Leslie Thomas's bawdy bestseller about the sexual adventures of British national servicemen in Singapore in the 1950s. Obvious jokes, lively playing, good sense of period.
RCA/Columbia VHS, Beta R

A Wedding*
US 1978 121m ● colour

Carol Burnett, Paul Dooley, Amy Stryker, Mia Farrow, Peggy Ann Garner
D, Robert Altman

A fashionable wedding party in the Midwest gives director Altman the framework for a kaleidoscopic examination of American social mores, skilfully juggling a large assortment of characters and giving strong injections of black humour.
CBS/Fox VHS, Beta A

That Sinking Feeling

The Virgin Soldiers

What's New Pussycat?

What's New Pussycat?
US/France 1965 108m ● colour

Peter O'Toole, Peter Sellers, Capucine,
Paula Prentiss, Romy Schneider
D, Clive Donner

Sex farce with more pace than wit. O'Toole is a magazine editor having problems with beautiful girls and finding Sellers' mad German psychiatrist little help. The writer, well below his best form, was Woody Allen.
Warner   VHS, Beta   A

What's Up Doc?*
US   1972   87m   v   colour

Barbra Streisand, Ryan O'Neal, Kenneth
Mars, Austin Pendleton, Sorrell Booke
D, Peter Bogdanovich

The madcap adventures of a scatty musicologist and a girl drifter in San Francisco. Highly popular, but overrated, pastiche of the Hollywood screwball comedy of the 1930s with clever individual gags but little overall style.
Warner   VHS, Beta   A

Whisky Galore***
GB   1949   80m   v   bw

Basil Radford, Joan Greenwood,
Catherine Lacey, Bruce Seton
D, Alexander Mackendrick

A sharp, witty rendering, with excellent characterisation and striking location photography, of Compton Mackenzie's novel about a battle of wits between Hebridean islanders and the customs man over a load of shipwrecked whisky.
Thorn EMI   VHS, Beta   B

Young Frankenstein*
US   1974   106m   v   bw

Gene Wilder, Peter Boyle, Marty
Feldman, Madeline Kahn
D, Mel Brooks

Brooks' parody of the classic horror genre, appropriately filmed in steely black and white, is the usual infuriating ragbag of jokes that work and jokes that fail; but worth enduring the corn for the odd moments of brilliance.
CBS/Fox   VHS, Beta   A

Drama

ENCOUN

RANK
VIDEO
LIBRARY

Absence of Malice
US 1981 116m v colour

Paul Newman, Sally Field, Bob Balaban,
Melinda Dillon, Luther Adler
D, Sydney Pollack

Field as a hard-nosed reporter investigating the disappearance of a union boss, Newman oddly cast as the liquor merchant who becomes a prime suspect. Well aimed attack on journalistic abuse but compromised by its glossiness.
RCA/Columbia VHS, Beta, CED A

Accident**
GB 1967 100m v colour

Dirk Bogarde, Stanley Baker, Jacqueline
Sassard, Michael York, Vivien Merchant
D, Joseph Losey

Fine performances by Bogarde and Baker as Oxford dons whose private lives are laid bare by an Austrian girl student. Harold Pinter's screenplay is a masterly example of conveying far more than the words spoken.
Thorn EMI VHS, Beta A

The African Queen***
GB 1951 101m v colour

Humphrey Bogart, Katharine Hepburn,
Robert Morley, Peter Bull, Theodore Bikel
D, John Huston

Chalk-and-cheese encounter on a Congo tugboat during the First World War between the rough, hard-drinking skipper and a prim missionary. Enjoyment all the way, with Bogie and Hepburn in scintillating form as the odd couple.
CBS/Fox CED only C

Aguirre, Wrath of God**
W Germany 1972 90m v col

Klaus Kinski, Cecilia Rivera, Roy Guerra,
Helena Rojo, Del Negro
D, Werner Herzog

Spanish conquistadors, under the self-styled "Wrath of God", go in search in El Dorado, the city of gold, with deadly results. A compelling and memorable study of megalomania which builds powerfully.
Palace VHS, Beta A

Alice Doesn't Live Here Anymore*
US 1975 108m v colour
Ellen Burstyn, Alfred Lutter, Kris Kristofferson, Billy Green Bush
D, Martin Scorsese

The funny/sad adventures of a widow who takes to the road with her 11-year-old son hoping to realise her childhood dream of becoming a singer. Burstyn's Alice deservedly won the best actress Oscar.
Warner VHS, Beta A

Alice in the Cities*
W Germany 1974 110m v bw
Rudiger Vogeler, Yella Rottländer, Elisabeth Kreuzer, Edda Köchl
D, Wim Wenders

A German reporter, returning to Europe from an assignment in the United States, finds himself the unwilling guardian of a little girl who has lost contact with her family. Observant and amusing study of identity and alienation.
Palace VHS, Beta A

All the President's Men **
US 1976 130m v colour
Robert Redford, Dustin Hoffman, Jason Robards, Martin Balsam, Hal Holbrook
D, Alan J Pakula

Redford and Hoffman as the *Washington Post* reporters, Woodward and Bernstein, uncovering the White House link with the Watergate break-in which led to the downfall of President Nixon. Meticulous, authentic and as gripping as a thriller.
Warner VHS, Beta A

American Gigolo**
US 1980 114m ● colour
Richard Gere, Lauren Hutton, Hector Elizondo, Nina van Pallandt, Bill Duke
D, Paul Schrader

Gere convinces as an escort to wealthy women in Los Angeles. He is framed for a murder and a client he was with refuses to give him an alibi. Cool, detached and absorbing.
CIC VHS, Beta A

Animal Farm
GB 1954 73m v colour
Voices: Maurice Denham, Gordon Heath
D, John Halas/Joy Batchelor

Cartoon treatment of George Orwell's brilliant satire on Soviet Russia, a fable about animals who overthrow their master and erect a new tyranny in his place. Stays closely to the story but misses its biting wit.
Rank VHS, Beta B

Anna Karenina
GB 1947 110m v bw

*Vivien Leigh, Keiron Moore, Ralph
Richardson, Marie Lohr
D, Julien Duvivier*

Decorative but dull version of Tolstoy's
novel of a Russian aristocrat's wife who
destroys herself in an affair with a young
cavalry officer. Richardson excellent as the
husband; Leigh leaves you longing for the
Garbo version.
Spectrum   VHS, Beta   B

Another Time, Another Place
GB 1983 101m v colour

*Phyllis Logan, Giovanni Mauriello,
Denise Coffey
D, Michale Radford*

A frustrated young farmer's wife in the
Scottish highlands has a guilty affair with
one of the Italian prisoners billeted on the
farm during the Second World War. Low-
key treatment of emotional dislocation.
VCL   VHS, Beta   A

Another Way*
Hungary 1982 105m ● colour

*Jadwiga Jankowska-Cieslak, Grazyna
Szapolowska, Jozef Kroner, Gabor
Revickky
D, Karoly Makk*

A lesbian relationship between two journalists in Hungary set against the political
tensions of the period following the 1956
uprising. Sensitive handling of a controversial theme with immaculate acting.
Palace   VHS, Beta   A

Ashes and Diamonds***
Poland 1958 100m v bw

*Zbigniew Cybulski, Ewa Krzyanowska,
Adam Pawlikowski
D, Andrzej Wajda*

The third and finest part of Wajda's trilogy
about the agony of Poland during the
Second World War. Cybulski as a young
partisan with a mission to kill a visiting
Communist leader on the eve of peace (see
also *A Generation* and *Kanal*).
Thorn EMI   VHS, Beta   C

Atlantic City**
Canada/France 1980 105m v col
*Burt Lancaster, Susan Sarandon, Kate
Reid, Michel Piccoli, Hollis McLaren
D, Louis Malle*

Lancaster giving one of his finest screen
performances as an old gangster living in
the past who gets caught up with drug
pushers. A powerfully melancholic piece,
to which the city itself forms an appropriately seedy backdrop.
Home Video Productions   VHS, Beta   A

Autobiography of a Princess**
GB   1975   60m   v   colour
*James Mason, Madhur Jaffrey
D, James Ivory*

Nostalgic memories of a vanished India are
stirred by the annual visit of Mason's
retired English tutor to the daughter of his
old employer, an Indian princess living in
London. A small film but exquisitely done.
With *The Delhi Way* (40m).
Virgin   VHS, Beta   A

Autumn Sonata**
Sweden/WG   1979   89m   v   col
*Ingrid Bergman, Liv Ullman, Halvar Bjork
D, Ingmar Bergman*

In the only screen collaboration between
the two Bergmans, Ingrid plays a concert
pianist visiting her daughter after a long
interval and becoming involved in a painful
reappraisal of their past relationship.
Precision   VHS, Beta   B

Bad Timing**
GB   1980   118m   ● colour
*Art Garfunkel, Theresa Russell, Harvey
Keitel, Denholm Elliott, Daniel Massey
D, Nicolas Roeg*

Many-layered examination, told mainly in
flashback, of a psychoanalyst's obsessive
affair with an American girl in Vienna
which nearly ends in tragedy. Raw and
vibrant, a virtuoso display of cinema
technique.
Rank   VHS, Beta   A

Badlands**
US   1973   90m   ● colour
*Martin Sheen, Sissy Spacek, Warren
Oates, Ramon Bieri
D, Terrence Malick*

Remarkable first feature from Malick,
displaying great narrative skill and feel for
landscape, about the doomed odyssey
across rural America of a teenage girl and a
young garbage collector who turns killer.
Warner   VHS, Beta   A

The Ballad of Joe Hill*
Sweden 1971 100m v colour
Thommy Berggren, Anja Schmidt, Evert Anderson, Cathy Smith
D, Bo Widerberg
The true story of a young Swede who emigrates to America just after the turn of the century and becomes a martyr to the emerging trade union movement. Lyrical, low-key treatment with nicely observed scenes of immigrant life.
Thorn EMI   VHS, Beta   A

Barry Lyndon*
GB 1975 184m v colour
Ryan O'Neal, Marisa Berenson, Patrick Magee, Hardy Kruger
D, Stanley Kubrick
Elegantly crafted adaptation of the novel by Thackeray about the adventures of an eighteenth century Irish gentleman. The human interest takes second place to the ravishing photography, which is inevitably diminished on the small screen.
Warner   VHS, Beta   A

Becket
GB 1964 142m v colour
Richard Burton, Peter O'Toole, John Gielgud, Martita Hunt, Sian Phillips
D, Peter Glenville
Handsomely staged, unimaginative and over-talkative historical drama from a play by Jean Anouilh about the rift between Henry II (O'Toole) and Becket (Burton) which ends in the latter's assassination in Canterbury Cathedral.
Odyssey   VHS, Beta   A

La Belle et La Bête**
France 1946 86m v bw
Jean Marais, Josette Day, Mila Parely, Marcel André, Michel Auclair
D, Jean Cocteau
Extraordinary personal re-working of the fairytale of beauty and the beast, full of visual surprises and stylistic experiments. Marais adds an extra dimension of ambiguity by being cast as both the beast and the villainous Avenant.
Thorn EMI   VHS, Beta   C

Betrayal*
GB 1982 74m v colour
Jeremy Irons, Ben Kingsley, Patricia Hodge, Avril Elgar
D, David Jones
Coolly efficient rendering of Harold Pinter's play which explores the complex relationships between a publisher, his wife and her former lover, starting in the present day and making several shifts back in time.
Virgin   VHS, Beta   A

Alice in the Cities

Bad Timing

The Ballad of Joe Hill

Bicycle Thieves***
Italy 1948 90m v bw

Lamberto Maggiorani, Enzo Staiola
D, Vittorio de Sica

The most celebrated example of Italian neo-realism, shot on the streets and using non-professional actors, in which an unemployed man and his small son search Rome for the stolen bicycle on which his job depends.
Longman VHS, Beta B

A Bill of Divorcement*
US 1932 75m v bw

John Barrymore, Katharine Hepburn,
Billie Burke, David Manners
D, George Cukor

Notable for Katharine Hepburn's impressive first film performance as a girl desperately in love but reluctant to marry for fear of inheriting her father's insanity. From a play by Clemence Dane.
Guild VHS, Beta A

The Birth of a Nation***
US 1915 158m v bw

Henry B Walthall, Mae Marsh, Ralph
Lewis, Lillian Gish, George Siegman
D, D W Griffith

Epic of the American Civil War which interweaves the stories of two families, one from the north and the other from the south. Melodramatic and racialist, maybe, but a huge step forward in the evolution of the cinema as a new art.
Spectrum VHS, Beta A

The Birthday Party*
GB 1968 122m v colour

Robert Shaw, Patrick Magee, Sidney
Tafler, Dandy Nichols, Moultrie Kelsall
D, William Friedkin

Useful film record of Harold Pinter's first full-length play, a "comedy of menace" in which the hapless Stanley, a lodger in a seaside boarding house, is interrogated and humiliated by two mysterious strangers.
Rank VHS, Beta A

The Bitter Tears of Petra von Kant*

W Germany 1975 124m ● col

Hanna Schygulla, Margit Carstensen, Irm Hermann
D, Rainer Werner Fassbinder

Claustrophobic and stylised study of jealousy and betrayal, shot on a single set, and exploring the obsessive relationships of a woman fashion designer with her passive assistant and an aspiring model.
Palace   VHS, Beta   A

Black Narcissus**

GB 1946 100m ● colour

Deborah Kerr, David Farrar, Flora Robson, Kathleen Byron, Esmond Knight
D, Michael Powell/Emeric Pressburger

A characteristically unconventional offering from the Powell–Pressburger team about the sexual and other deprivations of Anglo-Catholic nuns in the Himalayas. Stunning to look at and, remarkably, shot entirely in the studio.
Rank   VHS, Beta   B

The Blood of a Poet**

France 1930 58m v bw

Lee Miller, Pauline Carton, Odette Talazac, Enrique Rivero
D, Jean Cocteau

Allegorical fantasy which makes dramatic use of the resources of the cinema, particularly its ability to dissolve the barriers of time and space, to explore a favourite Cocteau theme, the nature of artistic inspiration.
Palace   VHS, Beta   A

Blow Up*

GB 1966 104m ● colour

David Hemmings, Sarah Miles, Vanessa Redgrave, Peter Bowles, John Castle
D, Michelangelo Antonioni

Puzzle-piece, set in the London of the "swinging sixties", in which a young fashion photographer thinks he has discovered a murder only to have the body disappear. That rather sums up the film itself – less than meets the eye.
MGM/UA   VHS, Beta   A

Blue Collar*

US 1978 109m v colour

Richard Pryor, Harvey Keitel, Yaphet Kotto, Ed Begley Jnr, Harry Bellaver
D, Paul Schrader

Discontented by poor pay and a bullying foreman, three workers in a Detroit car plant set off on a half serious plan to steal the union funds and get themselves embroiled in blackmail, corruption and murder.
CIC   VHS, Beta, V2000   A

The Blue Lagoon
US 1980 98m ● colour

Brooke Shields, Christopher Atkins, Leo McKern, William Daniels
D, Randal Kleiser

H de Vere Stacpoole's Victorian novel of the two children shipwrecked on a tropical island, grow up to become lovers and have a child. Self-consciously modern treatment of a tale of innocence; pretty to look at.
RCA/Columbia VHS, Beta A

Breaker Morant*
Australia 1980 106m v colour

Edward Woodward, Jack Thompson, John Waters, Bryan Brown, Charles Tingwell
D, Bruce Beresford

Courtroom drama about three Australian army officers on trial for murdering prisoners during the Boer War. An attempt to win sympathy for the accused and expose the hypocrisy of the accusers.
Guild VHS, Beta A

Breakfast at Tiffany's*
US 1961 109m v colour

Audrey Hepburn, George Peppard, Patricia Neal, Martin Balsam
D, Blake Edwards

The bitter-sweet relationship between a zany New York playgirl, Holly Golightly, and a struggling young writer. Cleaned-up version of Truman Capote's novel, featuring the Henry Mancini-Johnny Mercer song hit, "Moon River".
CIC VHS, Beta B

Breaking Away*
US 1979 96m v colour

Dennis Christopher, Dennis Quaid, Daniel Stern, Jackie Earle Haley, Barbara Barrie
D, Peter Yates

Affectionate study of small town America which follows four Indiana teenagers as they adjust to the adult world after leaving high school and take on the local university students in a cycle race on the campus.
CBS/Fox VHS, Beta, Laser A

Breaking Glass
GB 1980 104m v colour

Hazel O'Connor, Jon Finch, Phil Daniels, Jonathan Pryce
D, Brian Gibson

Gritty tale of a pop band and its reluctant lead singer, who finds fame too much to cope with. Fresh performances and an attempt to get to grips with contemporary youth culture help to compensate for a hackneyed plot.
VCL VHS, Beta C

Brief Encounter***
GB   1945   86m   v   bw

*Celia Johnson, Trevor Howard, Stanley
Holloway, Joyce Carey, Cyril Raymond
D, David Lean*

Railway station romance between a suburban housewife and a married doctor
which both know they cannot sustain. A
beautifully crafted and impeccably acted
film, with a lot more emotional truth than
some later critics have allowed.
Rank   VHS, Beta   A

Brimstone and Treacle
GB   1982   87m   ●   colour

*Sting, Denholm Elliot, Joan Plowright,
Suzanna Hamilton, Benjamin Whitrow
D, Richard Loncraine*

From a Dennis Potter play, which gained
notoriety by being banned by the BBC,
about the strange young visitor to a
suburban household who stirs feelings of
guilt over the crippled daughter as a
prelude to raping her.
Brent Walker   VHS/Beta, V2000   A

Britannia Hospital*
GB   1982   111m   v   colour

*Malcolm McDowell, Leonard Rossiter,
Graham Crowden, Joan Plowright
D, Lindsay Anderson*

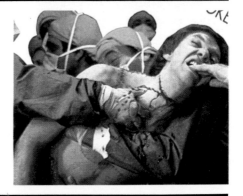

A strike-torn hospital awaiting a royal visit
is an acid metaphor for contemporary
Britain. Savaged by most critics and a box-
office disaster, it may yet emerge as the
masterpiece that some people already
think it is.
Thorn EMI   VHS, Beta   A

Caesar and Cleopatra*
GB   1945   121m   v   colour

*Claude Rains, Vivien Leigh, Stewart
Granger, Flora Robson, Cecil Parker
D, Gabriel Pascal*

A grandiose screen version of the George
Bernard Shaw play which took two and a
half years to make and became the most
expensive British film. Its troubled production history, more than its artistic quality,
lends a certain fascination.
Rank   VHS, Beta   A

Can She Bake a Cherry Pie?*
US   1983   90m   v   colour

*Karen Black, Michael Emil, Michael
Margotta, Frances Fisher
D, Henry Jaglom*

She has been abandoned by her husband;
he is divorced and a shoulder to cry on; and
after false starts a relationship develops. A
slight tale, but told with much perception
and humour; two fine performances.
Virgin   VHS, Beta   A

The Candidate*
US 1972 106m v colour
Robert Redford, Don Porter, Melvyn Douglas, Karen Carlson
D, Michael Ritchie

Redford as an idealistic young lawyer who is persuaded to run for the Senate and is unexpectedly successful, but at the cost of his marriage and his principles. A convincingly jaundiced view of American politics.
Warner VHS, Beta A

Carnal Knowledge*
US 1971 97m ● colour
Jack Nicholson, Art Garfunkel, Candice Bergen, Ann-Margret, Rita Moreno
D, Mike Nichols

Follows the mainly unsuccessful sex lives of two college friends from student days into middle age, with Nicholson and Garfunkel as the contrasting pair. Has a dated air and not as daring as it seemed; but splendid acting all round.
CBS/Fox VHS, Beta, CED, Laser A

Casablanca***
US 1942 102m v bw
Humphrey Bogart, Ingrid Bergman, Paul Henreid, Claude Rains, Conrad Veidt
D, Michael Curtiz

What started as just another efficiently crafted Hollywood melodrama was a big commercial success and picked up three Oscars. It has matured over the years into a cult, one of the handful of films that can happily be seen over and over again. The ingredients are a meaty story; a nicely balanced and perfectly orchestrated cast in which even the smaller parts are filled by actors of the distinction of Sydney Greenstreet, Peter Lorre and John Qualen; and the technical polish, particularly the atmospheric lighting and camerawork, of a Hollywood studio (Warner Brothers) at the peak of its powers. What also keeps the film fresh is the nostalgia it evokes for that period early in the Second World War when Europe fought alone and North Africa was an escape route for refugees from the Nazis. Bogart plays a man caught up in this traffic, the owner of Rick's Bar in Casablanca; he melts his cynicism sufficiently to help an old flame (Bergman) escape with her husband (Henreid) to the United States. For Bogart and Bergman *Casablanca* provided arguably their most famous film roles, yet neither, curiously, was a first choice. We can only wonder George Raft, Dennis Morgan or Ronald Reagan—yes!—might have made of Rick, or how adequately Michèle Morgan or Hedy Lamarr would have filled the Bergman role. The film's much quoted line, "Play it again, Sam", does not appear; what Bergman actually said was "Play it, Sam", a reference to the song *As Time Goes By*, which is performed by the pianist Dooley Wilson.
Warner VHS, Beta, CED A

Casanova
Italy 1976 163m ● colour
*Donald Sutherland, Tina Aumont, Cicely
Brown, Carmen Scarpitta
D, Federico Fellini*
Episodic, rambling treatment of the sexual
adventures of the famous 18th century
lecher which is often striking visually but
dramatically disappointing. The hero's trail
of seductions is not particularly erotic.
CBS/Fox   VHS, Beta   A

Cat on a Hot Tin Roof*
US 1958 105m ● colour
*Paul Newman, Burl Ives, Elizabeth
Taylor, Jack Carson, Judith Anderson
D, Richard Brooks*
From the Tennessee Williams play which
lifts the lid on the emotional tangles of a
rich Southern family. Barely disguises its
stage origin but strongly acted by Newman
(the impotent son), Taylor (his suffering
wife) and Ives (Big Daddy).
MGM/UA   VHS, Beta, Laser   B

The Champ
US 1979 117m v colour
*Jon Voight, Faye Dunaway, Ricky
Schroder, Jack Warden, Arthur Hill
D, Franco Zeffirelli*
Handkerchiefs at the ready for this third
screen version of the old tearjerker about
the drunken, washed-up boxer idolised by
his small son. Appealing performance by
young Schroder and lyrical photography.
MGM/UA   VHS, Beta, V2000, Laser   A

Champion**
US 1949 98m v bw
*Kirk Douglas, Arthur Kennedy, Marion
Maxwell, Paul Stewart
D, Mark Robson*
Douglas in his first starring role as an
unscrupulous boxer sacrificing family and
friends on his way to the top. Stylish and
tightly directed exposé of the fight game,
and still one of the best boxing films.
Spectrum   VHS, Beta   B

The Chant of Jimmy Blacksmith
Australia 1978 122m ● colour
*Tommy Lewis, Jack Thompson, Freddy
Reynolds, Ray Barrett, Angela Punch
D, Fred Schepisi*
The tragedy of a half-caste aborigine who
becomes a victim of racial bigotry in
Australia at the turn of the century and is
goaded into murderous revenge. Powerful
plea for tolerance.
Odyssey   VHS, Beta   A

Chariots of Fire**
GB 1981 121m □ colour
*Ben Cross, Ian Charleson, Nigel Havers,
Cheryl Campbell, Ian Holm
D, Hugh Hudson*
Oscar-winning study of the motives which
drive an ambitious Jew and a devout Scot
to run for Britain in the 1924 Olympics.
Very adroit piece of cinema, with nice feel
for period, strong characters and a rattling
good story.
CBS/Fox   VHS, Beta, V2000, Laser   A

Charly*
US 1968 106m ● colour

*Cliff Robertson, Claire Bloom, Leon
Janney, Lilia Skala
D, Ralph Nelson*

The story of a mentally retarded bakery
worker whose condition is transformed by
surgery. Robertson's sympathetic, Oscar-
winning performance is often at odds with
director Nelson's weakness for sentimentality.
Rank VHS, Beta A

The China Syndrome*
US 1979 120m v colour

*Jane Fonda, Jack Lemmon, Michael
Douglas, Scott Brady, Peter Donat
D, James Bridges*

A compelling exposé/thriller. Fonda is a
reporter who discovers a near-accident at a
nuclear power plant and comes up against
a conspiracy by the authorities to suppress
her story.
RCA/Columbia VHS, Beta A

The Cincinnati Kid*
US 1965 101m v colour

*Steve McQueen, Edward G Robinson,
Karl Malden, Tuesday Weld, Ann-
Margret
D, Norman Jewison*

Young card shark McQueen and old master
Robinson square up for a memorable battle
of wits across the poker table. Director
Jewison deftly builds the tension; Ann-
Margret and Tuesday Weld provide
unnecessary romantic interest.
MGM/UA VHS, Beta A

Class
GB 1983 98m ● colour

*Rob Lowe, Jacqueline Bisset, Andrew
McCarthy, Stuart Margolin
D, Lewis John Carlino*

A shy 19-year-old college boy has a
passionate affair with an older woman who
turns out to be the mother of his best friend.
A muddled treatment of the adolescence-
into-manhood theme, uncertain how
seriously to take itself.
Rank VHS, Beta A

Come Back to the Five and Dime Jimmy Dean, Jimmy Dean*

US 1982 105m ● colour

Sandy Dennis, Cher, Karen Black
D, Robert Altman

A reunion of a James Dean fan club at a "five and dime" store where memories of the women's idol are mixed with a look back on their own lives. Bold attempt to film a one-set play.
Intervision VHS, Beta A

Coming Home*

US 1978 90m v colour

Jon Voight, Jane Fonda, Bruce Dern, Robert Carradine
D, Hal Ashby

A Vietnam film which concentrates on the people at home. Fonda is the lonely wife who has an affair with a paralysed veteran (Voight) while her husband (Dern) is away fighting. Old-fashioned romanticism with superior acting.
Intervision VHS, Beta A

Conduct Unbecoming

GB 1975 102m v colour

Michael York, Stacy Keach, Trevor Howard, Susannah York, Richard Attenborough
D, Michael Anderson

Flatly handled version of the West End stage hit about a young subaltern in nineteenth century India who is wrongly accused of assaulting the widow of the former regimental captain.
Thorn EMI VHS, Beta A

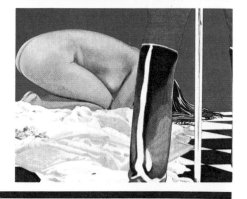

The Conversation*

US 1974 109m v colour

Gene Hackman, John Cazale, Allen Garfield, Frederic Forrest
D, Francis Ford Coppola

Superbly self-effacing performance by Hackman as a professional surveillance expert who bugs an apparently innocent conversation, develops a conscience about what he is doing and is drawn into a personal nightmare.
Arena VHS, Beta, V2000 B

Conversation Piece
Italy/France   1974   120m   v   col
Burt Lancaster, Helmut Berger, Claudia
Marsani, Silvana Mangano
D, Luchino Visconti

Echoes of Visconti's earlier *Death in Venice* (qv) in this sombre study of a reclusive professor persuaded to let his top floor flat to a group of young people whose presence causes him to review the disappointments of his life.
VCL   VHS, Beta   A

Cool Hand Luke*
US   1967   115m   v   colour
Paul Newman, George Kennedy, Jo Van
Fleet, J D Cannon, Lou Antonio
D, Stuart Rosenberg

Virtuoso performance from Newman as the member of a chain gang who becomes a hero to his fellow prisoners while the guards try to break him. Can be taken as a Christ allegory; certainly an astute and stylish piece of cinema.
Warner   VHS, Beta   A

The Damned*
WG/Italy   1969   148m   ●   col
Dirk Bogarde, Ingrid Thulin, Helmut
Berger, Renaud Verley, Helmut Griem
D, Luchino Visconti

Overblown and melodramatic account of how personal conflicts, internal power struggles and political divisions destroy a family of German steel tycoons as the Nazis tighten their grip on the country. Oozes with decadence.
Warner   VHS, Beta   A

Dark Victory*
US   1939   102m   v   bw
Bette Davis, George Brent, Humphrey
Bogart, Ronald Reagan
D, Edmund Goulding

Davis in one of her most famous tearjerking roles, as the spoiled society girl who discovers she is dying of a brain tumour and faces up to her end with courage. Unlikely casting for Bogart as an Irish groom.
Warner   VHS, Beta   A

Day for Night*
France/Italy   1973   116m   v   colour
Jean-Pierre Leaud, Jacqueline Bisset,
Valentina Cortese, Jean-Pierre Aumont,
François Truffaut
D, François Truffaut

Amusing, lightweight piece about the tribulations of a harassed film director (played by Truffaut) as his movie is held up by the squabbles and jealousies of the cast and beset by technical problems.
Warner   VHS, Beta   A

Days of Heaven**
US 1978 91m v colour

*Richard Gere, Brooke Adams, Sam
Shepard, Linda Manz
D, Terrence Malick*

Three young people leave Chicago for the
wheatfields of Texas, where an interlude of
idyllic happiness is the prelude to tragedy.
Superbly photographed (an Oscar for
cameraman Nestor Almendros) evocation
of place and period (1916).
CIC VHS, Beta A

Death in Venice**
Italy 1971 125m ● colour

*Dirk Bogarde, Bjorn Andresen, Silvana
Mangano, Marisa Berenson, Mark Burns
D, Luchino Visconti*

A German composer's obsession with a
handsome boy in a Venice succumbing to
cholera on the eve of the First World War.
Atmospheric piece, enriched by the beauty
of the images and the haunting music by
Gustav Mahler.
Warner VHS, Beta A

Deep End*
W Germany 1974 88m ● col

*Jane Asher, John Moulder Brown, Diana
Dors, Karl Michael Vogler
D, Jerzy Skolimowski*

Touching and often funny account of a
teenage boy's sexual awakening at a
municipal baths, as he pursues an obsessive relationship with the attractive female
attendant. Set in London but mainly shot in
Munich.
Capstan VHS, Beta A

The Deer Hunter***
US 1978 176m ● colour

*Robert de Niro, Christopher Walken,
Meryl Streep, John Cazale, John Savage
D, Michael Cimino*

A powerful statement about the effect of
the Vietnam war on three steelworkers
from Pennsylvania. It does not preach, but
contains intricate set pieces – a wedding, a
deer shoot, a game of Russian roulette –
which speak for themselves.
Thorn EMI VHS, Beta A

Deliverance*
US 1972 100m ● colour

*Burt Reynolds, Jon Voight, Ned Beatty,
Ronny Cox, James Dickey
D, John Boorman*

Gripping allegory about four city dwellers who set out on a test of manhood by canoeing down a dangerous river. They discover, ironically, that the main threat to their survival is themselves, not nature.
Warner VHS, Beta A

Despair*
WG/France 1978 108m v col

*Dirk Bogarde, Andrea Ferreol, Klaus
Löwitsch, Volker Spengler
D, Rainer Werner Fassbinder*

Adaptation by Tom Stoppard of the Nabokov novel about the disintegrating business and personal life of a Russian emigré chocolate manufacturer in Berlin in 1930. As sombre as it sounds but adeptly made; convincing central performance.
Thorn EMI VHS, Beta A

The Devils
GB 1970 105m ● colour

*Vanessa Redgrave, Oliver Reed, Dudley
Sutton, Max Adrian, Gemma Jones
D, Ken Russell*

Russell going right over the top with an hysterical, tasteless and often incoherent piece about sexually possessed nuns in seventeenth century France. The director's energy and visual flair give it a macabre fascination.
Warner VHS, Beta A

Doctor Zhivago*
US 1965 186m v colour

*Omar Sharif, Julie Christie, Rod Steiger,
Geraldine Chaplin, Alec Guinness
D, David Lean*

Meticulously crafted, strikingly photographed but weakly cast and dramatically staid version of Pasternak's great novel about the doomed love of an idealistic Russian doctor against the background of the 1917 Revolution.
MGM/UA VHS, Beta, V2000 A

Dog Day Afternoon*
US 1975 120m ● colour

*Al Pacino, John Cazale, Charles Durning,
James Broderick, Chris Sarandon
D, Sidney Lumet*

Based on a true incident of three men holding up a Brooklyn bank to steal money for a sex-change operation; two of them get trapped inside. Authentic locations, sympathetic performances and laced with a desperate humour.
Warner VHS, Beta A

A Doll's House
GB/France 1973 102m v col

*Jane Fonda, David Warner, Trevor
Howard, Edward Fox, Delphine Seyrig
D, Joseph Losey*

Ibsen's play about a wife who finally breaks free from a loveless marriage given contemporary feminist interest by the casting of Fonda as the heroine. But Howard's playing takes the honours, amid suitably chilly Norwegian locations.
Rank VHS, Beta A

Cat on a Hot Tin Roof

Despair

Dr Zhivago

Don't Look Now**
GB 1973 106m ● colour

Donald Sutherland, Julie Christie, Hilary
Mason, Clelia Matania, Massimo Serrato
D, Nicolas Roeg

A densely textured piece, full of allusive
meanings, about a couple who lose their
small daughter in a drowning accident but
are offered hope by a medium. Based on
the Daphne du Maurier story and shot
mainly in a wintry Venice.
Thorn EMI   VHS, Beta   A

The Dove
US 1974 100m v colour

Joseph Bottoms, Deborah Raffin, John
McLiam, Dabney Coleman
D, Charles Jarrott

Based on the true story of a 16-year-old
yachtsman, Robin Lee Graham, setting out
on a round the world voyage and finding
rough seas and romance on the way. By far
the best part is the photography of Sven
Nykvist.
Thorn EMI   VHS, Beta   A

The Draughtsman's Contract*
GB 1982 106m v colour

Anthony Higgins, Janet Suzman, Anne
Louise Lambert, Neil Cunningham
D, Peter Greenaway

Acclaimed low-budget British feature, set
in a country house in the 1690s. A young
draughtsman agrees to supply a set of
drawings in return for the lady's bed.
Elegant, stylised and constantly probing.
Guild   VHS, Beta   A

The Duellists
GB 1977 101m v colour

Keith Carradine, Harvey Keitel, Albert
Finney, Edward Fox, Tom Conti
D, Ridley Scott

Carradine and Keitel as two hussar officers
in the Napoleonic army who satisfy their
honour by engaging in a series of duels
over 16 years. Thin material for a feature
film but the photography is vivid and the
fights well staged.
CIC   VHS, Beta   A

Easy Rider**

US 1969 94m ● colour

*Peter Fonda, Dennis Hopper, Terry
Southern, Jack Nicholson
D, Dennis Hopper*

Enormously influential low-budget movie
with many imitators about two motor cycle
drop-outs whose disenchanted journey
across America on their Harley Davidson
choppers becomes a protest against a
repressive society.
RCA/Columbia   VHS, Beta, CED   A

The Electric Horseman

US 1979 120m v colour

*Robert Redford, Jane Fonda, Valerie
Perrine, Willie Nelson
D, Sydney Pollack*

Far-fetched fable about a disgruntled ex
rodeo champion (Redford) who has
descended to advertising breakfast cereals
and rides into the desert on his sponsor's
horse. Fonda is the television reporter who
follows him to get a story.
CIC   VHS, Beta, Laser   A

The Elephant Man**

US 1980 119m v bw

*John Hurt, Anthony Hopkins, Anne
Bancroft, John Gielgud, Wendy Hiller
D, David Lynch*

The extraordinary true story of a hideously
deformed sideshow freak in Victorian
London, who is rescued by a young
surgeon and becomes the talk of society.
Cleverly suggestive portrait by Hurt behind
the grotesque make-up.
Thorn EMI   VHS, Beta   A

Elmer Gantry*

US 1960 140m v colour

*Burt Lancaster, Jean Simmons, Arthur
Kennedy, Shirley Jones
D, Richard Brooks*

Highly effective, Oscar-winning performance by Lancaster, cynically turning
evangelism to commercial advantage in
the American midwest in the 1920s.
Simmons as the exploited healer, Jones
the girl friend turned hooker.
Warner   VHS, Beta   A

Endless Love
US   1981   110m   v   colour
Brooke Shields, Martin Hewitt, Shirley Knight, Don Murray
D, Franco Zeffirelli
Shades of Romeo and Juliet as a Chicago high school student's passion for a 15-year-old girl runs into family opposition. Shallow, sentimental and predictable with lots of lingering close-ups.
Spectrum   VHS, Beta, V2000, Laser   A

Les Enfants du Paradis***
France   1945   180m   v   bw
Arletty, Jean-Louis Barrault, Pierre Brasseur, Marcel Herrand
D, Marcel Carné
One of the masterworks of the cinema, a story of love and jealousy set in the stylishly evoked Parisian street theatre of the 1840s. Haunting performances by Barrault, as the mime, and Arletty, the elusive object of his passion.
Thorn EMI   VHS, Beta   C

The Enigma of Kaspar Hauser*
W Germany   1974   110m   ●   col
Bruno S, Hans Musaus, Henry Van Lyck, Enno Patalas
D, Werner Herzog
Unusual psychological drama based on the real case in nineteenth century Nuremberg of a 16-year-old boy who suddenly emerges into the world after being locked in a dark cellar all his life.
Palace   VHS, Beta   A

Equus
GB   1977   137m   ●   colour
Richard Burton, Peter Firth, Colin Blakely, Joan Plowright
D, Sidney Lumet
From Peter Shaffer's lauded play about a 17-year-old boy who has blinded six horses; a psychiatrist attempts to discover why. Less effective on film, which goes for realism and leaves little for the imagination.
Warner   VHS, Beta   A

Escape from Alcatraz
US   1979   112m   v   colour
Clint Eastwood, Patrick McGoohan,
Roberts Blossom, Paul Benjamin
D, Don Siegel

Eastwood as Frank Morris, a man who set
out in 1960 to do the impossible and break
out of the impregnable island prison in San
Francisco bay. Less an action film than a
careful dissection of the conventions of
prison life.
CIC   VHS, Beta, Laser   A

The Europeans*
GB   1979   83m   v   colour
Lee Remick, Robin Ellis, Tim Woodward,
Wesley Addy, Lisa Eichhorn
D, James Ivory

Subtle, low-key adaptation of the Henry
James novel about a European baroness
arriving in Boston to meet her wealthy
American cousins and find herself a
husband. Misses some of the Jamesian
irony but flawlessly acted.
Virgin   VHS, Beta   A

Excalibur
US   1981   140m   v   colour
Nicol Williamson, Helen Mirren, Nicholas
Clay, Nigel Terry, Cherie Lunghi
D, John Boorman

A reworking of the legend of King Arthur
and the knights of the Round Table which
dutifully covers familiar ground while
showing signs of looking for a contemporary idiom. Often enticing to watch, but
palls as drama.
Warner   VHS, Beta   A

The Executioner's Song
US   1982   135m   ●   colour
Tommy Lee Jones, Rosanna Arquette, Eli
Wallach, Christine Lahti
D, Lawrence Schiller

The last months of the life of the murderer
Gary Gilmore, executed in 1977 after
opposing all attempts to get the sentence
reduced. Scripted, from his book, by Norman Mailer; strong on the facts, weak on
character.
Palace   VHS, Beta   A

The Fallen Idol***
GB   1948   91m   v   bw
Ralph Richardson, Michèle Morgan,
Bobby Henrey, Sonia Dresdel, Jack
Hawkins
D, Carol Reed

A touching study of misguided loyalty,
written by Graham Greene. A small boy in
a London embassy lies to protect his friend,
the butler, who is suspected of killing his
wife. A near-perfect essay in miniature.
Thorn EMI   VHS, Beta   C

Family Life

GB 1971 102m v colour

Sandy Ratcliff, Bill Dean, Grace Cave
D, Ken Loach

Searching examination, using the ultra-realist techniques evolved by Loach for television, of a 19-year-old girl's mental breakdown and the part played by unsympathetic parents and the failure of psychiatric treatment.

Thorn EMI VHS, Beta C

Far from the Madding Crowd*

GB 1967 159m v colour

Julie Christie, Peter Finch, Alan Bates, Terence Stamp, Prunella Ransome
D, John Schlesinger

Ambitious version of Hardy's tragic novel which captures the look and feel of his Victorian Wessex but falls short on the human side. Christie leaves the acting honours to Finch's Boldwood.

Thorn EMI VHS, Beta A

The Farmer's Daughter*

US 1940 97m v bw

Loretta Young, Joseph Cotten, Ethel Barrymore, Charles Bickford
D, H C Potter

Loretta Young gives an Oscar-winning performance in a light-hearted tale of a determined Swedish farm girl from Minnesota who wins a seat in Congress and finds romance. An enjoyable trifle, glossily mounted in the old Hollywood style.

Guild VHS, Beta A

Fear Eats the Soul*

WG 1973 93m v colour

Brigitte Mira, El Hedi Ben Salem, Barbara Valentin, Irm Hermann
D, Rainer Werner Fassbinder

A fable of modern Germany in which an unlikely romance between a young Moroccan immigrant worker and a 60-year-old charlady blossoms into marriage but is undermined by the social prejudices of her family and friends.

Palace VHS, Beta A

Fedora**

WG/France   1978   108m   v   col

William Holden, Marthe Keller, Hildegard Knef, José Ferrer
D, Billy Wilder

Holden as a Hollywood producer trying to lure an old star out of retirement and discovering the secret of her eternal beauty. A mellower treatment of image-making previously explored in Wilder's *Sunset Boulevard.*
Thorn EMI   VHS, Beta   A

Fellini Satyricon

Italy   1969   124m   ●   colour

Martin Potter, Hiram Keller, Salvo Randone, Max Born
D, Federico Fellini

Visually powerful but fragmented account of a student's sexual odyssey through ancient Rome in which the director seems more concerned to indulge his private fantasies than to reach out to the audience.
Warner   VHS, Beta   A

F.I.S.T.

US   1978   133m   v   colour

Sylvester Stallone, Rod Steiger, Peter Boyle, Melinda Dillon, David Huffman
D, Norman Jewison

Jimmy Hoffa was the model for this biopic of an idealistic union man who becomes corrupted by power and brings about his own downfall. Stallone's rough charm helps to make the early parts convincing; the rest is melodrama.
Warner   VHS, Beta   A

Fitzcarraldo**

WG   1982   160m   v   colour

Klaus Kinski, Claudia Cardinale, Jose Lewgoy, Paul Hittscher
D, Werner Herzog

Original and engrossing story of an Irishman's eccentric ambition to build his own opera house in the Peruvian jungle. Contains the splendidly bizarre episode of a steamboat being moved from one river to another across a mountain.
Palace   VHS, Beta   A

Five Days One Summer

US   1982   108m   v   colour

Sean Connery, Betsy Brantley, Lambert Wilson, Jennifer Hilary, Isabel Dean
D, Fred Zinnemann

Lyrical Alpine photography and Zinnemann's seasoned craftsmanship cannot disguise the thinness of this triangle drama. Set in the 1930s, it involves a middle-aged Scottish doctor, his young mistress and a handsome Swiss guide.
Warner   VHS, Beta   A

For Whom the Bell Tolls*

US   1943   128m   v   colour

Gary Cooper, Ingrid Bergman, Akim Tamiroff, Katina Paxinou
D, Sam Wood

Romance between an American schoolteacher and an orphan girl against the background of the Spanish Civil War. Reverential Hollywood treatment of the Hemingway novel, with notable performances from the two stars.
CIC   VHS, Beta   B

Fox and His Friends*

WG   1975   123m   ● colour

Rainer Werner Fassbinder, Karl-Heinz Boehm, Peter Chatl, Harry Bär
D, Rainer Werner Fassbinder

A piece of irony, for the main contribution of Fox's friends is to swindle and desert him. Played by Fassbinder, he is a homosexual sideshow performer whose life gradually falls apart after he loses his job and his lover.
Palace   VHS, Beta   A

Frances

US   1982   133m   v   colour

Jessica Lange, Sam Shepard, Kim Stanley
D, Graeme Clifford

Story of a Hollywood victim, the actress Frances Farmer, who takes to drugs and is committed to an asylum. A film unsure where to place the blame but held together by the anguished performance of Jessica Lange.
Thorn EMI   VHS, Beta   A

The French Lieutenant's Woman*

GB   1981   121m   v   colour

Jeremy Irons, Meryl Streep, Leo McKern, Patience Collier, Hilton McRae
D, Karel Reisz

John Fowles' complex novel about a Victorian gentleman's passion for a woman apparently jilted by her French lover was intriguingly adapted by Harold Pinter. Much to admire, a few reservations.
Warner   VHS, Beta, CED   A

From the Life of the Marionettes*

WG   1980   99m   ● colour

Robert Atzorn, Christine Buchnegger, Martin Benrath
D, Ingmar Bergman

A typically bleak Bergman piece, fascinating in its way, about a young German businessman whose motive for strangling a prostitute is traced back to mental instability and a failed marriage.
Precision   VHS, Beta   A

The Fugitive*
US 1947 86m v bw

Henry Fonda, Dolores del Rio, Pedro
Armendariz, Ward Bond
D, John Ford

Fonda as a tormented priest on the run
from an anti-clerical regime in a highly
symbolic, visually ravishing version of
Graham Greene's book *The Power and the
Glory*. With *Follow Me Quietly*, a 1949
thriller.
Kingston   VHS, Beta   A

The Games
GB 1969 95m □ colour

Stanley Baker, Michael Crawford, Ryan
O'Neal, Charles Aznavour
D, Michael Winner

Four long-distance runners from different
countries prepare for the Olympic marathon. The build-up is less interesting than
the race itself, which is convincingly staged
and genuinely exciting.
CBS/Fox   VHS, Beta, Laser   A

Gandhi**
GB 1982 180m v colour

Ben Kingsley, Candice Bergen, Martin
Sheen, Edward Fox, Ian Charleson
D, Richard Attenborough

Oscar-scooping biopic of the non-violent
campaigner for Indian independence.
Honestly crafted, sumptuously photographed and with an exceptional performance from Kingsley, who convincingly
ages over more than half a century.
RCA/Columbia   VHS, Beta   A

The Garden of Allah*
US 1936 85m v colour

Marlene Dietrich, Charles Boyer, Basil
Rathbone, Tilly Losch, C Aubrey Smith
D, Richard Boleslawski

Dietrich as a socialite finding romance in
the Algerian desert with a Trappist monk
who has broken his vows. A rich slice of
Hollywood hokum, with two stars at the top
of their form, and an interesting early
example of Technicolor.
Guild   VHS, Beta   A

A Generation**
Poland   1954   86m   v   bw

Tadeusz Lomnicki, Urszula Mordzynska,
Roman Polanski, Zbigniew Cybulski
D, Andrzej Wajda

A graphic study of young Poles achieving personal and political maturity by joining the resistance movement against the German occupation of their country during the Second World War. The first of the Wajda "trilogy".
Thorn EMI   VHS, Beta   C

Gentleman Jim*
US   1942   104m   v   bw

Errol Flynn, Alan Hale, Alexis Smith,
John Loder, Ward Bond
D, Raoul Walsh

Amiable, undemanding biopic of the boxer James J Corbett, a world heavyweight champion of the 1890s. He was the first man to win the title under the Marquess of Queensberry Rules which demanded the use of gloves.
Warner   CED   C

Georgia*
US   1981   113m   v   colour

Craig Watson, Jodi Thelen, Michael
Huddleston, Jim Metzler
D, Arthur Penn

The relationships between four friends — three boys and Georgia, the girl they all love — as they grow up in America in the early 1960s. Interesting attempt at a Hollywood art movie.
Rank   VHS, Beta   A

The Girl with Green Eyes*
GB   1963   89m   v   bw

Peter Finch, Rita Tushingham, Lynn
Redgrave
D, Desmond Davis

An Irish country girl's brief affair with a middle-aged writer in Dublin; adaptation by Edna O'Brien of her novel, helped along by Tushingham's wide-eyed innocence, a nicely understated performance by Finch and pleasing locations.
Thorn EMI   VHS, Beta   B

The Go-Between**
GB   1970   112m   v   colour

Alan Bates, Julie Christie, Michael
Redgrave, Dominic Guard, Margaret
Leighton
D, Joseph Losey

L P Hartley's novel of a 12-year-old boy's innocent involvement in a secret love affair in late Victorian England. Expertly adapted for the cinema with a precise Harold Pinter script and evocative sense of period.
Thorn EMI   VHS, Beta   A

Gone With the Wind***

US   1939   220m   v   colour

Clark Gable, Vivien Leigh, Olivia de Havilland, Leslie Howard, Thomas Mitchell, Hattie McDaniel

D, *Victor Fleming*

The making of *Gone With the Wind* is an epic story in itself: how the producer, David O Selznick, overcame his initial doubts to buy the rights to Margaret Mitchell's best-selling novel of the American Civil War; how he made a special deal with MGM for the release of their contract star, Clark Gable, to play the key role of Rhett Butler; and his much publicised search for an actress to play Scarlett O'Hara, which ended with the choice of the inexperienced Vivien Leigh in preference to many better-known names. As for directors, the film

started with George Cukor, who was replaced by Victor Fleming (who had to leave another project, *The Wizard of Oz);* when Fleming fell ill, Sam Wood was engaged as back-up and is said to be responsible for just over 30 minutes of the finished film. But this is a producer's picture, not a director's, and its enormous commercial success — the biggest grossing film in cinema history until overtaken by *The Sound of Music* in the 1960s — is due largely to Selznick's showmanship and flair for spectacle, of which the burning of Atlanta is the oustanding example. A strong romantic story certainly helped; and the casting was uniformly successful. Altogether, it won nine Oscars, then a record.

MGM/UA   VHS, Beta   B

Great Expectations***
GB 1946 118m v bw

John Mills, Finlay Currie, Martita Hunt,
Jean Simmons, Alec Guinness
D, David Lean

Charles Dickens' novel brought triumphantly to the screen, with a host of fine
performances led by Mills' Pip; Oscar-
winning camerawork (Guy Green) and art
direction (John Bryan); and the fine narrative skill of the director, Lean.
Rank VHS, Beta A

The Great White Hope*
US 1970 100m v colour

James Earl Jones, Jane Alexander, Lou
Gilbert, Joel Fluellen
D, Martin Ritt

Thinly disguised biopic of Jack Johnson,
who became the first black boxer to win the
world heavyweight championship. Uncertain treatment of the racial aspect but held
together by James Earl Jones' mercurial
portrait of the fighter.
CBS/Fox VHS, Beta, Laser A

Hamlet**
GB 1948 155m v bw

Laurence Olivier, Eileen Herlie, Basil
Sydney, Jean Simmons
D, Laurence Olivier

Despite a considerable shortening of the
text and some over-indulgent camerawork,
this is a powerful rendering of Shakespeare's great tragedy; strong on atmosphere and with a compelling central
performance.
Rank VHS, Beta A

Finding a film
Notes on some of the ten categories in
this book.

Action/adventure: war, swashbucklers, historical epics, disaster
movies, martial arts; broad drama.

Comedy: slapstick silents, black
comedy (e.g. *Cul de Sac*), parodies (for
instance, *Cat Ballou* and *Blazing
Saddles* appear in this section, not
under Westerns).

Drama: human and domestic stories
rather than global ones.

Horror: witchcraft, monsters, supernatural. Hitchcock's *Psycho* and *The
Birds* are in this section because their
effect is horror-inducing.

Science fiction/fantasy; the unreal
world, especially the imagined world of
a different time (e.g. *Superman* and *Star*

Thrillers; gangsters, suspense, police,
domestic crime and spies.

If you still cannot find what you are
after, there is an index to all the titles at
the end, plus indexes on the main
directors and stars.

Hardcore*

US 1978 108m ● colour

George C Scott, Peter Boyle, Season Hubley, Dick Sargent
D, Paul Schrader

Scott on the trail of an errant daughter who has rejected her strict Calvinist upbringing to take part in pornographic films. Responsible treatment of a lurid theme, marred by a tendency to caricature the two milieus.
RCA/Columbia VHS, Beta A

Heartland

US 1979 95m v colour

Conchata Ferrell, Rip Torn, Barry Primus, Lilia Skala
D, Richard Pearce

Conscientious reconstruction, from documents relating to a true case, of the toils and struggles of a woman housekeeper at a remote ranch in Wyoming in the early part of the century. Comes persuasively close to reality.
Thorn EMI VHS, Beta A

Heat and Dust*

GB 1982 130m v colour

Julie Christie, Shashi Kapoor, Christopher Cazenove, Madhur Jaffrey, Greta Scacchi
D, James Ivory

A two-pronged look at the British experience in India, with Christie retracing the steps of her great-aunt who came to the country as a young bride in the 1920s. Ivory gently explores the contrasts.
3M VHS, Beta, V2000 A

Heatwave

Australia 1981 95m v colour

Judy Davis, Richard Moir, Chris Haywood, Bill Hunter
D, Phillip Noyce

Opposition by community activists to a new apartment block during a Christmas heatwave in Sydney takes a sinister turn when one of the protesters disappears and a man dies in a fire. Quasi-documentary drama, vividly presented.
Guild VHS, Beta, V2000 A

Henry V***

GB 1944 137m v colour

Laurence Olivier, Robert Newton, Leslie Banks, Esmond Knight
D, Laurence Olivier

Olivier as the young English king doing battle with the French in his brilliant adaptation of Shakespeare's play. It successfully combines cinematic spectacle with respect for the text; a distinguished score by William Walton.
Rank VHS, Beta A

Hobson's Choice**
GB 1953 102m v bw
*Charles Laughton, Brenda de Banzie,
John Mills, Richard Wattis, Helen Haye
D, David Lean*

Laughton in fine blustering form as the Lancashire bootmaker who gets his come-uppance from his wilful daughter and her meek husband; faithful and effective version of the famous play, if slightly too theatrical as comedy/drama.
Thorn EMI VHS, Beta C

The Householder*
India 1963 101m v bw
*Shashi Kapoor, Leela Naidu
D, James Ivory*

The first feature film collaboration between director Ivory, producer Ismail Merchant and writer Ruth Prawer Jhabvala. An engaging account of a young Indian school-teacher's attempt to come to terms with his arranged marriage.
Virgin VHS, Beta A

Hullabaloo over Georgie and Bonnie's Pictures*
India/GB 1978 82m v colour
*Peggy Ashcroft, Victor Banerjee, Saeed Jaffrey, Larry Pine, Aparna Sen
D, James Ivory*

The battle of wits between an American collector and the buyer for a London museum for a priceless collection of paintings belonging to an Indian maharajah. Witty, perceptive and gently ironic.
Virgin VHS, Beta A

The Hunted
GB 1970 90m ● colour
*Robert Shaw, Malcolm McDowell
D, Joseph Losey*

Nightmarish study of two men on the run who are pursued by a helicopter and soldiers as they make for a friendly border. Technically accomplished but dramatically dull allegory, the wider meaning of which remains unclear. Released in the cinema as *Figures in a Landscape*.
CBS/Fox VHS, Beta A

I Never Promised You a Rose Garden
US 1977 96m ● colour
*Kathleen Quinlan, Bibi Andersson, Sylvia Sidney, Ben Piazza
D, Anthony Page*

The painful path to rehabilitation of a 16-year-old girl who is committed to a psychiatric hospital after attempting suicide. Well-intentioned social drama but glossily presented and slushy.
Warner VHS, Beta A

I'll Be Seeing You*
US 1944 85m v bw
*Joseph Cotten, Ginger Rogers, Spring Byington, Tom Tully, Shirley Temple
D, William Dieterle*

Sentimental romance, in which a wounded soldier on Christmas leave (Cotten) falls in love with a convicted killer (Rogers), delivered in the lush Hollywood style. Good example of wartime fatalism with a famous title song.
Guild VHS, Beta B

The Go-Between.

Hardcore

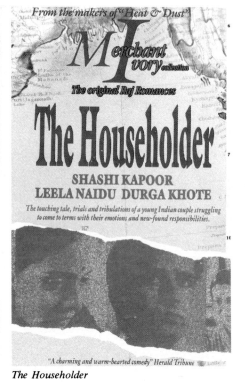

The Householder

Images*
Eire 1972 98m ● colour

Susannah York, René Auberjonois,
Marcel Bozzuffi, Hugh Millais
D, Robert Altman

An enigmatic study of a women's obsession with her past lovers which is either emptily pretentious or brilliantly perceptive, depending on your point of view. Certainly it is absorbing and enhanced by the stunning Irish locations.
VCL   VHS, Beta   C

Inserts
GB 1975 114m ● colour

Richard Dreyfuss, Jessica Harper,
Veronica Cartwright, Bob Hoskins
D, John Byrum

A cynical look at the Hollywood dream factory of the 1930s. A fading silent star and a once-acclaimed director, reduced to making blue movies, seek consolation in drugs. Tends to parade what it is trying to condemn.
Warner   VHS, Beta   A

Inside Moves
US 1980 109m v colour

John Savage, Diana Scarwid, David
Morse, Amy Wright
D, Richard Donner

Attempt to establish disability as a legitimate subject for commercial cinema. It charts the friendship among a group of handicapped people who frequent a Californian bar. But it leaves the audience unsure whether to laugh or cry.
Intervision   VHS, Beta   A

Interiors*
US 1978 91m v colour

Kirstin Griffith, Marybeth Hurt, Diane
Keaton, Richard Jordan, E G Marshall
D, Woody Allen

Could have been sub-titled "Woody Allen meets Ingmar Bergman". This intense and introverted analysis of the emotional problems of a wealthy New York lawyer and his family is right in the Bergman mould; and indulgent with it.
Warner   VHS, Beta   A

Intermezzo*
US 1939 70m v bw

Leslie Howard, Ingrid Bergman, John
Halliday, Edna Best, Cecil Kellaway
D, Gregory Ratoff

A sublime cinema weepie, with Bergman as the young pianist falling in love with her married teacher. It was the original version of this story, filmed in Sweden, that brought her to Hollywood and launched her international career.
Guild   VHS, Beta   A

Liquid Sky

Images

Intolerance***
US 1916 113m □ bw

Mae Marsh, Lillian Gish, Constance Talmadge, Robert Harron
D, D W Griffith

Four linked stories on the theme of man's intolerance, from ancient Babylon to modern America; an astonishing achievement for its time and like *Birth of a Nation* (qv) marking a giant step forward in the evolution of film art.
Spectrum VHS, Beta B

Jane Austen in Manhattan*
GB/US 1980 108m v colour

Anne Baxter, Robert Powell, Michael Wager, Tim Chaote, John Guerrasio
D, James Ivory

A delightfully witty and perceptive account of how two New York theatre companies come up with contrasting interpretations of *Sir Charles Grandison*, a rediscovered play by the young Jane Austen.
Virgin VHS, Beta A

Joan of Arc
US 1948 145m v colour

Ingrid Bergman, Jose Ferrer, George Coulouris, Francis L Sullivan, Gene Lockhart
D, Victor Fleming

A lush but studio-bound rendering of the Maid of Orleans. Of interest chiefly as an example of how Hollywood could take a noble theme and pulp it into just another big star vehicle.
MEVC VHS, Beta B

Le Jour Se Lève***
France 1939 85m v bw

Jean Gabin, Jules Berry, Arletty, Jacqueline Laurent
D, Marcel Carné

The fatalistic account of the last hours of a man who has killed his rival in love and barricades himself in his attic room as the police lay siege. An archetypal example, still very potent, of the French school of "poetic realism".
Virgin VHS, Beta A

Jubilee*
GB 1978 103m ● colour

Jenny Runacre, Little Nell, Toyah Willcox, Jordan, Adam Ant
D, Derek Jarman

A black vision of England, past, present and future, in which Queen Elizabeth I is transported into the late twentieth century and a society where law and order has given way to punk rule. Extravagant and provocative.
VCL VHS, Beta A

Julia*

US 1977 113m v colour

Jane Fonda, Vanessa Redgrave, Jason Robards, Maximilian Schell, Hal Holbrook
D, Fred Zinnemann

Fonda as the writer Lillian Hellman, recalling the fight of her childhood friend (Redgrave) against fascism in Europe during the 1930s; Robards as Hellman's companion, Dashiell Hammett. Strong acting, discreet direction.
CBS/Fox VHS, Beta, V2000, Laser A

Junior Bonner**

US 1972 99m v colour

Steve McQueen, Robert Preston, Ida Lupino, Joe Don Baker, Barbara Leigh
D, Sam Peckinpah

Subtle, low-key, quietly effective work from a director more readily associated with blood and gore; an ageing rodeo star (McQueen) tries unsuccessfully to go back to his roots. Preston and Ida Lupino memorable as the parents.
Rank VHS, Beta A

Kanal**

Poland 1956 93m ● bw

Teresa Izewska, Tadeusz Janczar, Emil Kariewicz, Wienczylaw Glinski
D, Andrzej Wajda

The plight of Polish resistance workers, forced to take refuge in the sewers during the ill-fated rising against the German occupiers in 1944. A harrowing portrait of an oppressed people, slightly marred by sentimentality.
Thorn EMI VHS, Beta C

The Killing of Sister George*

US 1969 134m ● colour

Beryl Reid, Susannah York, Coral Browne, Ronald Fraser, Hugh Paddick
D, Robert Aldrich

A coarsened version of a British stage hit, but Beryl Reid is still splendid as the lesbian actress whose career and private life collapses when she is dropped from a soap opera. Coral Browne equally good as her rival in love.
Rank VHS, Beta A

A Kind of Loving*
GB 1962 107m ● bw

Alan Bates, June Ritchie, Thora Hird,
Bert Palmer, Gwen Nelson
D, John Schlesinger

A British film very much of its period, when
working class anti-heroes and grimy northern landscapes were all the fashion.
Bates as a young draughtsman trapped
into a wretched marriage and trying to
make the best of it.
Thorn EMI VHS, Beta C

Knife in the Water**
Poland 1962 90m ● bw

Leon Niemczyk, Jolanta Umecka,
Zygmunt Malanowicz
D, Roman Polanski

Polanski's impressively assured first
feature explores the tensions and jealousies
which result when a young married couple
give a lift to a boy hitchhiker and invite him
to spend a weekend aboard their yacht.
Thorn EMI VHS, Beta C

Kotch*
US 1971 114m ● colour

Walter Matthau, Deborah Winters,
Felicia Farr, Charles Aidman
D, Jack Lemmon

Matthau in a heaven-sent role as a
crotchety old grandpa who helps an unwed
teenager during the birth of her baby.
Accomplished directorial debut by Lemmon, who largely avoids the sentimentality
inherent in his theme.
Guild VHS, Beta, V2000 A

Kramer vs Kramer*
US 1979 104m v colour

Dustin Hoffman, Justin Henry, Meryl
Streep, Jane Alexander, Howard Duff
D, Robert Benton

Oscar-garlanded and very popular piece
about an advertising executive bringing up
his young son after his wife leaves him and
his court battle to retain custody when she
wants the boy back. Slightly too cute to be
true.
RCA/Columbia VHS, Beta, CED A

Kuroneko*
Japan 1968 99m ● bw
*Kichiemon Nakamura, Nobuko Otowa,
Kiwako Taichi, Kai Sato
D, Kaneto Shindo*

Macabre and compelling fantasy set in medieval Japan with two women raped and murdered by samurai returning to terrorise the area in the guise of black cats. An unedifying tale, redeemed by its cinematic skill.
Palace VHS, Beta A

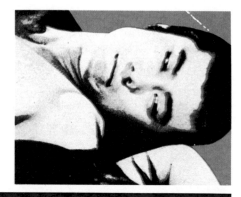

Kwaidan*
Japan 1964 164m ● colour
*Rentaro Mikuni, Ganemon Nakamura,
Katsuo Nakamura, Noburo Nakaya
D, Masaki Kobayashi*

Three Japanese ghost stories, notable for their striking images and ravishing colour. In the first a young samurai is haunted by his dead wife; in the second the reflection of a face in a teacup comes unexpectedly to life.
Palace VHS, Beta A

The Lacemaker
Fra/Switz/WG 1977 100m col
*Isabelle Huppert, Yves Benneyton,
Florence Giorgetti, Anne-Marie Düringer
D, Claude Goretta*

The tentative affair between a young Parisian hairdresser and the student she meets on a seaside holiday. It founders on differences of class and intellect. Carefully observed and understated drama with good locations.
VCL VHS, Beta A

Lady Caroline Lamb
GB 1972 118m v colour
*Sarah Miles, Jon Finch, Richard
Chamberlain, Margaret Leighton
D, Robert Bolt*

Handsomely mounted but misconceived tale of the English Regency. The capricious Lady Caroline embarks on a disastrous marriage with the politician Lord Melbourne but finds consolation in her pursuit of the poet Byron.
Thorn EMI VHS, Beta A

The Last Detail*
US 1973 100m ● colour

Jack Nicholson, Otis Young, Randy Quaid, Clifton James, Carol Kane
D, Hal Ashby

Nicholson and Young as a couple of tough navy officers given the job of escorting a listless 18-year-old to prison in New Hampshire and determined to live it up on the way. Crisply executed tour of American low life.
RCA/Columbia VHS, Beta A

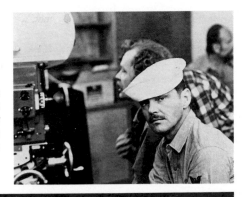

The Last Tycoon
US 1976 125m v colour

Robert de Niro, Robert Mitchum, Jack Nicholson, Ingrid Boulting
D, Elia Kazan

Scott Fitzgerald's story of a Hollywood mogul's obsessive pursuit of a girl who reminds him of his dead wife. A Harold Pinter script, some fine actors and a distinguished director should have produced something more memorable.
Videoform VHS, Beta A

The Last Valley
GB 1970 122m v colour

Michael Caine, Omar Sharif, Florinda Balkan, Nigel Davenport
D, James Clavell

A story of the Thirty Years War that is more a battle of minds than a battle of weapons. A ruthless mercenary leader (Caine) and a scholar (Sharif) meet in a peaceful valley that is untouched by the confict.
Rank VHS, Beta A

The Life and Death of Colonel Blimp**
GB 1943 163m v colour

Roger Livesey, Anton Walbrook, Deborah Kerr, Roland Culver
D, Michael Powell and Emeric Pressburger

A wonderfully irreverent treatment of the military mind (which is why Winston Churchill tried to have the film suppressed) in the story of a British soldier's adventures in three wars.
Rank VHS, Beta B

Lili Marleen
WG   1980   116m   v   colour
Hanna Schygulla, Giangarlo Giannini,
Mel Ferrer, Karl Heinz
D, Rainer Werner Fassbinder

The story of the German singer whose
recording of "Lili Marleen" makes her the
darling of the Nazis but compromises her
romance with a Jewish composer. A
curious and unsatisfactory nod by Fassbinder towards Hollywood melodrama.
Intervision   VHS, Beta, V2000   A

The Lion in Winter*
GB   1968   128m   v   colour
Katharine Hepburn, Peter O'Toole, Jane
Merrow, John Castle
D, Anthony Harvey

Extended war of words, during a Christmas
family reunion, between Henry II and his
wife, Eleanor of Aquitaine, over the succession to the throne. A strong piece of
theatre, dominated by two larger than life
performances.
Embassy   VHS, Beta, V2000   A

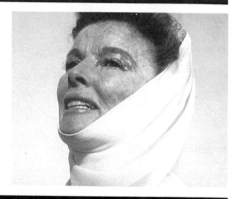

Liquid Sky
US   1982   107m   ●   colour
Anne Carlisle, Paula E Sheppard, Bob
Brady, Susan Doukas
D, Slava Tsukerman

Liquid sky is slang for heroin, which gives a
clue to this bizarre effort by Russian
expatriate film makers about the "new
wave" disco culture of young New Yorkers,
observed by the psychedelic eye of a flying
saucer.
VTC   VHS, Beta   A

Lisztomania
GB   1975   104m   ●   colour
Roger Daltrey, Sara Kestelman, Paul
Nicholas, Ringo Starr
D, Ken Russell

Only Russell would have come up with the
idea of turning the life of the composer Liszt
into a rock opera; and only he would have
produced this hysterical farrago of damp
parodies and blue jokes. But grant him his
cheek.
Warner   VHS, Beta   A

Lizards*
Italy   1963   85m   v   bw
Toni Petruzzi, Stefano Sattaflores, Sergio
Farrannino
D, Lina Wertmüller

An amused look at a small hillside town in
southern Italy, whose inhabitants are sunk
in an amiable lethargy and barely touched
by the world outside. Director Wertmüller
has a sharp eye for detail and does not try to
moralise.
Palace   VHS, Beta   A

Long Day's Journey into Night*
US 1961 135m ● bw

*Ralph Richardson, Katharine Hepburn,
Jason Robards, Dean Stockwell
D, Sidney Lumet*

Eugene O'Neill's mammoth compendium of raw-edged emotion, based on his own family. Director Lumet stays in the shadows and the actors respond magnificently; a fine example of filmed theatre.
Spectrum   VHS, Beta   B

Look Back in Anger*
GB 1959 95m ● colour

*Richard Burton, Mary Ure, Claire Bloom,
Edith Evans, Gary Raymond
D, Tony Richardson*

John Osborne's play exploded like a land-mine in the staid London theatre of the 1950s but does not have the same impact in this opened up film version. Burton as the anti-establishment hero, Jimmy Porter, declaims to effect.
Thorn EMI   VHS, Beta   B

Looking for Mr Goodbar
US 1977 136m ● colour

*Diane Keaton, Tuesday Weld, William
Atherton, Richard Kiley
D, Richard Brooks*

The double life of a young woman who teaches deaf children during the day and goes in search of casual sex by night, with horrifying consequences. Diane Keaton's game performance helps to leaven the sordidness of the theme.
CIC   VHS, Beta   A

Love Story*
US 1970 96m v colour

*Ali MacGraw, Ryan O'Neal, Ray Milland,
John Marley
D, Arthur Hiller*

It may not be a masterpiece but it was certainly a phenomenon, this smash hit tear jerker about an idyllic romance between two students which ends with her dying of cancer; worth a look, if only to see what the fuss was about.
CIC   VHS, Beta, Laser   A

The Loveless

US 1982 83m ● colour

Willem Dafoe, Robert Gordon, Martin
Kanter, J Don Ferguson
D, Kathryn Bigelow/Monty Montgomery

A brutal symphony of 1950s chrome and
leather as a motor cycle gang descend
upon a small southern town and become
involved in a family feud, which is pursued
to the death. Violent, charmless but with a
macabre sense of style.
Palace VHS, Beta A

Mahler*

GB 1974 115m ● colour

Robert Powell, Georgina Hale, Richard
Morant, Lee Montague
D, Ken Russell

Though suffering from typical Russell
extravagances and vulgarities, this is one
of his more coherent composer biopics.
Restrained performances from Powell as
Mahler and Hale as his wife; superb
photography and fine music.
Guild VHS, Beta, V2000 A

Making Love

US 1982 112m ● colour

Michael Ontkean, Kate Jackson, Harry
Hamlin, Wendy Hiller
D, Arthur Hiller

A married doctor leaves his wife for a gay
writer. Attempt to establish homosexuality
as an acceptable subject for mainstream
commercial cinema but too pat to be
convincing either as art or entertainment.
CBS/Fox VHS, Beta, V2000, Laser A

A Man Called Horse

US 1970 114m ● colour

Richard Harris, Judith Anderson, Jean
Gascon, Manu Tupou
D, Elliot Silverstein

The tale of an English aristocrat who in
1825 is taken prisoner by the Sioux
Indians, gains the respect of the tribe and
becomes their leader. Realistic, but dramatically uneven examination of Indian life
and customs.
CBS/Fox VHS, Beta B

A Man For All Seasons*

GB 1966 120m v colour

Paul Scofield, Wendy Hiller, Susannah
York, Robert Shaw, Orson Welles
D, Fred Zinnemann

Scofield makes a rare but telling film
appearance as Sir Thomas More, sacrificing his head for his principles. This
strongly cast and impeccably staged version of Robert Bolt's play picked up five
Oscars.
RCA/Columbia VHS, Beta A

Man Friday

GB 1975 104m A colour

Peter O'Toole, Richard Roundtree
D, Jack Gold

Adrian Mitchell's re-working of the classic
Defoe story. The slave, Friday, proves more
intelligent and resourceful than his master,
Robinson Crusoe. An ingenious idea but
ultimately unconvincing.
Precision VHS, Beta, V2000 A

Mandingo
US 1975 121m ● colour

James Mason, Susan George, Perry King, Richard Ward
D, Richard Fleischer

Brutal and unsympathetic account on life on a slave-breeding plantation in Lousiana in 1840; best not taken too seriously. The normally subtle Mason leads a roster of over-the-top performances.
CIC VHS, Beta, V2000 A

The Marriage of Maria Braun*
WG 1978 116m v colour

Hanna Schygulla, Klaus Lowitsch, Ivan Desny, Gottfried John
D, Rainer Werner Fassbinder

A German war bride has an affair with an American soldier after her husband is declared missing in action. Intermittently interesting piece in Fassbinder's spare style; overstretches the political allegory.
VCL VHS, Beta A

Mean Streets**
US 1973 110m ● colour

Harvey Keitel, Robert de Niro, David Proval, Amy Robinson
D, Martin Scorsese

Sharply etched picture of four young men of New York's "Little Italy", drawn from the director's own experiences. It underlines the brutality and aimlessness of the milieu as well as the strength of family ties.
Video Movies VHS, Beta A

Memoirs of a Survivor
GB 1981 111m ● colour

Julie Christie, Christopher Guard, Leonie Mellinger, Nigel Hawthorne
D, David Gladwell

Ambitious, but ultimately vain attempt to capture Doris Lessing's vision of a bleak urban future. Christie as the "survivor", a middle-aged woman who watches the violence and decay around her and seeks an alternative world.
Thorn EMI VHS, Beta A

Mephisto**
Hungary 1981 144m ● colour
*Klaus Maria Brandauer, Ildiko Bansaki,
Krystyna Janda
D, Istvan Szabo*

Critically acclaimed study of a pushy young
actor in the German theatre of the 1920s
who abandons his leftwing principles in
order to save his career under the Nazis.
Winner of the Oscar for best foreign film.
Palace   VHS, Beta   A

Merry Christmas Mr Lawrence*
GB/Japan 1983 120m ● col
*David Bowie, Ryuichi Sakamoto, Tom
Conti, Jack Thompson
D, Nagisa Oshima*

Set in a Japanese PoW camp exploring the
relationships between two British prisoners
and their captors. One (Bowie) wins their
admiration; the other (Conti) is caught
between the two.
Palace   VHS, Beta, V2000   A

Midnight Express
GB 1978 120m ● colour
*Brad Davis, Randy Quaid, John Hurt,
Irene Miracle, Bo Hopkins
D, Alan Parker*

Graphic indictment of Turkish prison atro-
cities, based on the story of an American
student, Billy Hayes, who spent five years
in the Sagamilcar jail for possessing
hashish. The effect is somewhat blunted by
overstatement.
RCA/Columbia   VHS, Beta, CED   R

The Misfits*
US 1961 110m v colour
*Clark Gable, Marilyn Monroe,
Montgomery Clift, Eli Wallach
D, John Huston*

Overly symbolic piece about modern-day
cowboys rounding up wild mustangs in the
Nevada desert, from a script by Arthur
Miller. The end of two celebrated careers;
Gable died before shooting was completed
and Monroe the following year.
Warner   VHS, Beta   A

Moby Dick*
GB 1956 110m v colour
*Gregory Peck, Richard Basehart, Harry
Andrews, Orson Welles
D, John Huston*

Intelligent, slow-moving adaptation of Her-
man Melville's novel, with Peck as a not too
convincing Captain Ahab, obsessively
seeking revenge on the great white whale
which has left him with only one leg.
Thorn EMI   VHS, Beta   A

Mommie Dearest

US   1981   124m   v   colour

Faye Dunaway, Diana Scarwid, Steve Forrest, Howard da Silva
D, Frank Perry

Biopic of Joan Crawford which stresses her cruelty to her children as recounted in the memoirs of her adopted daughter on which the film is based. Dunaway bears an uncanny physical likeness at times but conviction stops there.
CIC   VHS, Beta, V2000   A

Montenegro**

GB/Sweden   1981   87m   ●   col

Susan Anspach, Erland Josephson, Per Oscarsson, John Zacharias
D, Dusan Makavejev

From the director of the notorious *Mysteries of the Organism*, this is a tall tale about the bored wife of a rich Swedish businessman and her sexual adventures with a group of Yugoslav immigrants. Definitely for adults only.
Guild   VHS, Beta, V2000   A

Moonlighting*

GB   1982   97m   v   colour

Jeremy Irons, Eugene Lipinski, Jiri Stanislaw, Eugeniusz Haczkiewicz
D, Jerzy Skolimowski

The anxieties and tensions among four Polish building workers, as they renovate their boss's house in London while awaiting news of the Solidarity crisis back home. A perceptive study that transcends its topical context.
3M   VHS, Beta   A

The Music Lovers

GB   1970   118m   ●   colour

Richard Chamberlain, Glenda Jackson, Christopher Gable, Max Adrian
D, Ken Russell

Billed as a love affair between a homosexual and a nymphomaniac and not a bad description of Russell's treatment of the tormented life of Tchaikovsky. The usual Russell mixture of cinematic flair and shock for shock's sake.
Warner   VHS, Beta   A

My Brilliant Career*

Australia   1979   101m   v   colour

Judy Davis, Sam Neill, Wendy Hughes, Robert Grubb, Max Cullen
D, Gillian Armstrong

A much-praised performance by Judy Davis in a gently feminist piece about a girl growing up in the Australian bush at the turn of the century. She has to make the difficult choice between a safe marriage and independence.
Guild   VHS, Beta, V2000   A

Network*

US   1976   120m   v   colour

Peter Finch, William Holden, Faye Dunaway, Robert Duvall
D, Sidney Lumet

Paddy Chayevsky's pungent media satire with Finch (winning the first posthumous best actor Oscar) as a deranged television newscaster who ups his ratings by threatening to commit suicide on the air. Another Oscar went to Dunaway.
Warner   VHS, Beta   A

Mahler

A Man For All Seasons

Midnight Express

Newsfront*
Australia   1978   111m   v   bw/col

*Bill Hunter, Gerard Kennedy, Wendy
Hughes, Don Crosby
D, Philip Noyce*

Combination of fictional sequences and
newsreel footage which tell the social and
political history of Australia between 1948
and 1956. The friendly rivalry between two
newsreel companies is the dramatic peg.
Home Video Productions   VHS, Beta   A

Nicholas and Alexandra
GB   1971   189m   v   colour

*Michael Jayston, Janet Suzman,
Laurence Olivier, Jack Hawkins
D, Franklin J Schaffner*

Good looking, intelligently acted but ponderously handled slice of Russian history. It
covers the reign of the last Czar, Nicholas II,
from the early years of the century to his
overthrow in 1917 and execution.
RCA/Columbia   VHS, Beta   A

Nicholas Nickleby
GB   1947   107m   v   bw

*Derek Bond, Cedric Hardwicke, Alfred
Drayton, Sybil Thorndike
D, Alberto Cavalcanti*

The adventures of Dickens' Victorian
schoolteacher, deprived of his rightful
fortune by a wicked uncle. Polished craftsmanship that does not quite come alive,
despite several excellent character performances.
Thorn EMI   VHS, Beta   C

The Night Porter
Italy   1973   117m   ●   colour

*Dirk Bogarde, Charlotte Rampling,
Phillipe Leroy, Gabriele Ferzetti
D, Liliana Cavani*

A chance meeting in Vienna after the war
between a former SS officer and a girl who
was under his command in a concentration
camp leads to a resumption of their sado-
masochistic relationship. Uncomfortable
and unedifying.
Intervision   VHS, Beta, V2000   A

Norma Rae*

US 1979 109m v colour

Sally Field, Beau Bridges, Ron Leibman,
Pat Hingle, Barbara Baxley
D, Martin Ritt

Field winning the best actress Oscar as a tough southerner leading the fight for union recognition at the textile mill where she works. A serious-minded film, weakened by sentimentality and a tendency to blur the issues.
CBS/Fox   VHS, Beta, Laser   A

Now Voyager**

US 1942 115m v colour

Bette Davis, Claude Rains, Paul Henreid,
Gladys Cooper
D, Irving Rapper

Hollywood melodrama at its ripest with Davis as the neurotic young spinster who is cured by psychiatrist Rains and embarks on a doomed affair with Henreid. Max Steiner's lush score gained an Oscar and perfectly captures the mood.
Warner   VHS, Beta   A

O Lucky Man!*

GB 1973 167m ● colour

Malcolm McDowell, Arthur Lowe, Ralph
Richardson, Helen Mirren
D, Lindsay Anderson

A young coffee salesman's disenchanted journey through a Britain dominated by violence, corruption and repression. A long, episodic piece in the style of Voltaire's *Candide* and capable of several interpretations.
Warner   VHS, Beta   A

Odd Man Out**

GB 1946 113m v bw

James Mason, Robert Newton, Kathleen
Ryan, F J McCormick
D, Carol Reed

The last hours of a wounded gunman on the run in Belfast. A dazzlingly ornate film, with lighting and camerawork that skilfully captures the atmosphere of the city at night. A gallery of memorable characters.
Rank   VHS, Beta   A

An Officer and a Gentleman*

US 1982 126m ● colour

Richard Gere, Debra Winger, David
Keith, Louis Gossett Jr
D, Taylor Hackford

Troubled young man survives tough training course to become a Navy pilot and finds romance with a factory girl. Skilful mix of old-fashioned boy-makes-good and modern sexual frankness.
CIC   VHS, Beta   C

Old Boyfriends*

US 1978 103m ● colour

Talia Shire, Richard Jordan, John
Belushi, Keith Carradine
D, Joan Tewkesbury

After the failure of her marriage and an attempt to commit suicide, a woman makes a journey into the past through her former boyfriends in order to understand herself. Efficiently handled psychological case study.
Precision   VHS, Beta   A

Oliver Twist***
GB   1948   112m   ☐   bw

Alec Guinness, Robert Newton, John Howard Davies, Kay Walsh
D, David Lean

After *Great Expectations* (qv) Lean came up with another fine adaptation of Dickens. This polished piece of cinematic craftsmanship captures the flavour of the original and proceeds with great narrative skill.
Rank   VHS, Beta   A

On Golden Pond*
US   1981   105m   v   colour

Katharine Hepburn, Henry Fonda, Jane Fonda, Doug McKeon
D, Mark Rydell

Henry Fonda, in his last film, as a testy old college professor effecting a reconciliation with his daughter. A warm and sentimental story, with an idyllic New England setting, and an ideal showcase for its veteran stars.
Precision   All systems   A

On the Waterfront**
US   1954   108m   v   bw

Marlon Brando, Eva Marie Saint, Lee J Cobb, Rod Steiger
D, Elia Kazan

A powerful study of union corruption in the New York docks, with Brando as the young ex-boxer who decides to testify against the gang. Notable for many fine performances in the "Method" style and the stark dockside locations.
RCA/Columbia   VHS, Beta   A

One Flew Over the Cuckoo's Nest**
US   1975   129m   ●   colour

Jack Nicholson, Louise Fletcher, Brad Douris, William Redfield
D, Milos Forman

Tremendous acting from Nicholson as the newcomer to a mental asylum whose subversive behaviour brings him in conflict with the authorities. A very funny film but also a desperately sad one.
Thorn EMI   VHS, Beta   A

My Brilliant Career

Now Voyager

One From the Heart
US 1982 101m ● colour
*Frederic Forrest, Teri Garr, Nastassia
Kinski, Raul Julia
D, Francis Ford Coppola*

Coppola's grandiose and financially disastrous attempt to combine the lavish production values of the old Hollywood studios with a bitter-sweet romance in which the lovers part after a row and find themselves new partners.
CBS/Fox VHS, Beta A

Onibaba*
Japan 1964 105m ● colour
*Nobuko Otawa, Jitsuko Yoshimura, Kei
Sato
D, Kaneto Shindo*

Not for the squeamish. A mother and daughter-in-law survive in medieval Japan by killing exhausted soldiers and selling their armour. A deserter falls for the girl and sets off a triangle of sexual jealousy.
Palace VHS, Beta A

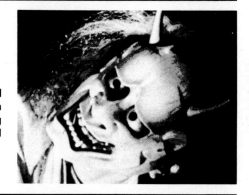

Ordinary People*
US 1980 119m ● colour
*Donald Sutherland, Mary Tyler Moore,
Timothy Hutton, Judd Hirsch
D, Robert Redford*

Intelligent and unsentimental study, which won an Oscar for Redford in his first attempt as a director. Wealthy Chicago parents, selfish mother and weak father, try to come to terms with their mentally disturbed son.
CIC VHS, Beta, V2000, Laser A

Orphée***
France 1949 90m v bw
*Jean Marais, François Périer, Maria
Casarès, Marie Déa
D, Jean Cocteau*

Cocteau's outstanding achievement in the cinema, a remarkable modern version of the legend of Orpheus seeking his wife, Eurydice, in the underworld. Here Orpheus is a famous poet who becomes fascinated by a mysterious princess.
Thorn EMI VHS, Beta C

The Other Side of Midnight
US 1977 159m ● colour

*Marie-France Pisier, John Beck, Susan
Sarandon, Raf Vallone
D, Charles Jarrott*

The on-off romance between a French girl
and an American airman is the strand
running through this lush concoction of
jealousy and revenge. It tries to ape the
classic Hollywood melodrama without the
same style.

CBS/Fox VHS, Beta, Laser, CED A

Out of Season
GB 1975 87m ● colour

*Cliff Robertson, Vanessa Redgrave,
Susan George, Edward Evans
D, Alan Bridges*

Robertson as an American returning to an
English seaside hotel, hoping after 20
years to revive a relationship that nearly led
to marriage; but the woman's daughter has
designs of her own. The acting almost
makes it credible.

Thorn EMI VHS, Beta A

Out of the Blue*
Canada 1980 89m ● colour

*Linda Manz, Dennis Hopper, Sharon
Farrell, Raymond Burr, Don Gordon
D, Dennis Hopper*

Terrifying study, not for the squeamish, of a
disturbed teenage girl's murderous revenge on her feckless parents. An explosive, uncompromising look at the punk
generation which has the virtue of stating
rather than judging.

CBS/Fox VHS, Beta A

Panic in Needle Park
US 1971 105m ● colour

*Al Pacino, Kitty Winn, Adam Vint,
Richard Bright, Kiel Martin
D, Jerry Schatzberg*

Unrelenting study of drug addiction, in
which Pacino is a small-time junkie who
befriends a young New Yorker and drags
her into the mire. An "awful warning" film
that tends to labour the point.

CBS/Fox VHS, Beta, Laser B

Papillon
US 1973 150m v colour

*Steve McQueen, Dustin Hoffman, Don
Gordon, Anthony Zerbe
D, Franklin Schaffner*

Despite its usually charismatic stars, this
story of two convicts – a wrongly convicted
murderer and a famous forger – trying to
escape from the brutal penal colony of
French Guyana remains curiously unin-
volving.

RCA/Columbia VHS, Beta A

The Passenger*
Ita/Fra/Sp 1975 118m ● col
Jack Nicholson, Maria Schneider, Jenny Runacre, Ian Hendry
D, Michelangelo Antonioni

A characteristic Antonioni puzzle piece, of which the viewer must make what he or she can. A television reporter (Nicholson) changes identity with a dead man and gets involved in gun-running for African revolutionaries.
MGM VHS, Beta B

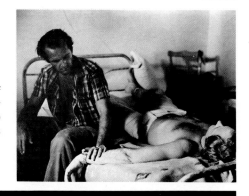

The Pawnbroker
US 1965 109m ● bw
Rod Steiger, Brock Peters, Geraldine Fitzgerald, Jaime Sanchez
D, Sidney Lumet

Steiger – unusually self-effacing as a Jewish pawnbroker in the Harlem slums – is tormented by the loss of his wife and children in a Nazi concentration camp. A well-intentioned and gruelling film that strives too hard for effect.
Spectrum VHS, Beta B

Performance*
GB 1970 101m ● colour
James Fox, Mick Jagger, Anita Pallenberg, Michele Breton
D, Nicolas Roeg and Donald Cammell

Original and unsettling. It starts as a gangster piece (Fox as a London thug bumping off the owner of a betting shop) and becomes an examination of identity as Fox, on the run, takes refuge with pop star Jagger and two girls.
Warner VHS, Beta A

Picnic at Hanging Rock*
Australia 1975 115m v colour
Rachel Roberts, Dominic Guard, Helen Morse
D, Peter Weir

Atmospheric and stylish. Set in Australia at the turn of the century, it concerns the mysterious disappearance of three girls and a teacher from a school party exploring a volcanic rock in the countryside.
Home Video Productions VHS, Beta A

The Picture Show Man*
Australia 1977 98m v colour
Rod Taylor, John Meillon, John Ewart, Harold Hopkins
D, John Power

The unassuming and highly enjoyable story of rival travelling cinema owners in the Australian outback in the 1920s, told with pace, humour and a nice sense of period. Taylor and Meillon play the protagonists.
Guild VHS, Beta A

An Officer and a Gentleman

The Ploughman's Lunch

Pixote
Brazil 1981 125m ● colour
Fernando Ramos da Silva
D, Hector Babenco
An eight-year-old boy from the slums of Sao Paulo gets initiated in the ways of crime at reform school and makes the rapid graduation from petty thief to murderer. Exposé of a social evil which overstates its case.
Palace VHS, Beta A

The Ploughman's Lunch
GB 1983 120m ● colour
Jonathan Pryce, Tim Curry, Rosemary Harris, Frank Finlay
D, Richard Eyre

Pryce as a cynical media man prepared to betray his roots and his principles, and manipulate his acquaintances, for his own self-seeking ends. Interesting mainly for its jaundiced view of British society in the 1980s.
Virgin VHS, Beta A

Popeye*
US 1981 92m □ colour
Robin Williams, Shelley Duvall, Ray Walston, Paul Dooley
D, Robert Altman

Popeye, a shipwrecked sailor, arrives at Sweethaven to find his long lost father and displaces the nasty Bluto in the affections of Olive Oyl. Interesting attempt to give flesh and blood to cartoon characters; little about spinach.
Disney VHS, Beta R

Les Portes de la Nuit*
France 1946 99m v bw
Pierre Brasseur, Yves Montand, Nathalie Nattier, Serge Reggiani
D, Marcel Carné
A stylised treatment of the problems of post-war France, derived from a ballet. The last full collaboration of director Carné and scriptwriter Jacques Prévert, which also produced *Les Enfants du Paradis* and *Le Jour Se Lève* (qqv).
Thorn EMI VHS, Beta C

Portrait of Jennie**
US 1948 86m v bw
Jennifer Jones, Joseph Cotten, Ethel Barrymore, David Wayne
D, William Dieterle
Cotten is a penniless artist who meets a girl in Central Park, becomes obsessed by her and paints her portrait only to discover she is the spirit of a woman who died long ago. Very likeable and expertly made fantasy.
Guild VHS, Beta A

Drama

Pretty Baby*
US 1978 106m ● colour

Keith Carradine, Susan Sarandon,
Brooke Shields, Francis Faye
D, Louis Malle

The sexual initiation of a 12-year-old girl, brought up in a New Orleans brothel by her prostitute mother. A theme that many found shocking but Malle's treatment is admirably restrained; he calls it a comedy. CIC VHS, Beta A

The Prime of Miss Jean Brodie*
GB 1969 99m v colour

Maggie Smith, Robert Stephens, Pamela
Franklin, Celia Johnson
D, Ronald Neame

Maggie Smith in her Oscar-winning role as the unconventional Edinburgh schoolmistress of the 1930s. Strongly acted but weakly directed, it misses the ironic flavour of its source, the novel by Muriel Spark. Guild VHS, Beta A

Prostitute
GB 1980 96m ● colour

Eleanor Forsythe, Kate Crutchley, Kim
Lockett, Nancy Samuels
D, Tony Garnett

Commendably unsensational look at a controversial subject, handled in a quasi-documentary style. A prostitute leaves Birmingham to try her luck in London, gets disillusioned and rejoins her old colleagues. Videospace VHS, Beta A

Quadrophenia
GB 1979 113m ● colour

Phil Daniels, Leslie Ash, Philip Davis,
Mark Wingett, Sting
D, Franc Roddam

A pungent examination of British youth culture in the 1960s: drugs, music and casual sex. It ends in a violent battle between Mods and Rockers on Brighton seafront. Based on an album by The Who, the film's producers. Spectrum VHS, Beta, V2000, Laser A

Quartet*
GB 1948 115m v bw

Basil Radford, Naunton Wayne, Mai Zetterling, Dirk Bogarde, George Cole D, Ralph Smart, Harold French, Arthur Crabtree, Ken Annakin

An enjoyable omnibus of four Somerset Maugham short stories — *The Facts of Life, The Alien Corn, The Kite* and *The Colonel's Lady* — adapted by R C Sherriff and introduced by Maugham himself.
Rank VHS, Beta A

The Queen of Spades**
GB 1948 90m v bw

Anton Walbrook, Edith Evans, Ronald Howard, Yvonne Mitchell D, Thorold Dickinson

Visually haunting film, it recalls the German silent cinema of the 1920s; from the Pushkin story about a Russian officer trying to wrest the secret of winning at cards from an old countess, a splendidly sinister Edith Evans.
Thorn EMI VHS, Beta C

Raggedy Man
US 1981 91m ● colour

Sissy Spacek, Eric Roberts, Sam Shepard, William Sanderson D, Jack Fisk

A curious tale about a woman bringing up her young sons in a Texan town during the Second World War. She is protected from two sadistic brothers by a mysterious odd-job man who is not quite what he seems.
CIC VHS, Beta A

The Raging Moon*
GB 1970 107m v colour

Malcolm McDowell, Nanette Newman, Georgia Brown, Bernard Lee D, Bryan Forbes

An unlikely love affair between two wheel-chair victims in a home for the physically disabled. Beautifully played by the principals, it avoids the twin pitfalls of morbidity and sentimentality.
Thorn EMI VHS, Beta A

Ragtime
US 1981 149m ● colour

James Olson, Mary Steenburgen, James Cagney, Pat O'Brien
D, Milos Forman

Fact-and-fiction panorama of America in the 1900s, focusing on a scandal over a chorus girl and a black jazz pianist's fight against white prejudice. Uneasy transcription of E L Doctorow's novel but nice to see Cagney again.
Thorn EMI   VHS, Beta   A

Rebecca***
US 1940 139m v bw

Laurence Olivier, Joan Fontaine, George Sanders, Judith Anderson
D, Alfred Hitchcock

Shy, vulnerable Joan Fontaine marries the handsome master of a Cornish mansion and is haunted by the unseen presence of his dead first wife. Atmospheric and gripping adaptation of the novel by Daphne du Maurier.
Guild   VHS, Beta   B

The Red Shoes***
GB 1948 132m □ colour

Anton Walbrook, Moira Shearer, Leonide Massine, Robert Helpmann
D, Michael Powell and Emeric Pressburger

The trite story of a young ballet dancer torn between love and art given glorious substance by the cinema's formal qualities of colour, decor and camerawork. Contains a complete ballet choreographed by Helpmann and Massine.
Rank   VHS, Beta   A

Reds*
US 1981 188m v colour

Warren Beatty, Diane Keaton, Jack Nicholson, Maureen Stapleton
D, Warren Beatty

Absorbing biopic of the American journalist, John Reed, who saw the Russian Revolution at first hand and wrote the classic book, *Ten Days That Shook the World*. A mammoth undertaking for Beatty, the star, director, producer and co-writer.
CIC   VHS, Beta, V2000   A

La Règle du Jeu***
France   1939   110m   v   bw

Marcel Dalio, Nora Gregor, Jean Renoir,
Mila Parély, Roland Toutain
D, Jean Renoir

Few would now dispute the claim of *La Règle du Jeu* (usually translated as *The Rules of the Game*) to be included in the half dozen or so greatest films ever made but this was not always the view. When it was first shown in Paris in July 1939 the audience jeered and even tried to burn down the cinema. The distributors demanded cuts to make the film more acceptable and the running time was reduced from 113 minutes to 85. In October 1939 the film was banned by the French Government as being morally unacceptable; and it was banned again by

the occupying Germans. After the war only mutilated versions remained and not until the 1950s, thanks to the chance discovery of missing fragments, were two young enthusiasts, under Renoir's supervision, able to restore *La Règle du Jeu* to its original length. But back to the beginning: why should *La Règle du Jeu*, which covers the events, comic and tragic, during a weekend country house party, have aroused such initial hostility? The answer has partly to do with the film's complexity and, more importantly, it was seen as an attack (only too well founded, as things turned out) on the feebleness of the French bourgeoisie. But *La Règle du Jeu* is not simply about people at one point in time; it is about humanity in general, as seen by one of the cinema's foremost artists.
Longman   VHS, Beta   B

Drama

Return of the Soldier
GB   1982   100m   v   colour

*Alan Bates, Glenda Jackson, Julie
Christie, Ann-Margret
D, Alan Bridges*

Bates plays a shell-shocked soldier who
comes back from the First World War and
finds difficulty picking up his old relationships, with his wife, his cousin and a
former sweetheart. Competent version of a
novel by Rebecca West.
Guild   VHS, Beta   A

Rich and Famous
US   1981   113m   ●   colour

*Jacqueline Bisset, Candice Bergen, David
Selby, Meg Ryan
D, George Cukor*

Previously filmed as *Old Acquaintance*, it is
the one about two college girls whose
friendship is later strained as one becomes
a highbrow novelist and the other a writer
of popular trash. Bette Davis and Miriam
Hopkins did it better.
MGM   VHS, Beta, V2000, CED   A

Ring of Bright Water
GB   1969   109m   □   colour

*Bill Travers, Virginia McKenna, Peter
Jeffrey, Roddy McMillan
D, Jack Couffer*

This likeable animal story is tailor-made for
family viewing. A dissatisfied civil servant
moves to a remote cottage in the Scottish
Highlands with his pet otter; from the best-
seller by Gavin Maxwell.
Guild   VHS, Beta   A

Key to symbols

□   suitable for all the family
v   vetting by parents advisable
●   adults only (mature teenagers
    may enjoy these films but
    parents should vet with care)

Note: Some films have been given a
'v' symbol although they are watched
and enjoyed by a wide range of age
groups (for example *Raiders of the Lost
Ark* in the Action/Adventure section
and the James Bond movies in Thrillers).

What the Stars mean

***   Not to be missed
**   Highly recommended
*   Well worth watching

Note: The absence of a star does not
mean that a film is worthless; rather, a
star denotes extra quality.

The Robe
US 1953 129m v colour
Richard Burton, Jean Simmons, Michael Rennie, Victor Mature
D, Henry Koster

Reverently handled Biblical epic with Burton as the Roman tribune in charge of the crucifixion but later converted to Christianity. The first film to be made in Cinemascope, though its impact is lessened on the small screen.
CBS/Fox   VHS, Beta   B

Robin and Marian
US 1976 107m v colour
Sean Connery, Audrey Hepburn, Robert Shaw, Ronnie Barker, Nicol Williamson
D, Richard Lester

Version of Robin Hood in which the swashbuckling is largely ignored in favour of a curious mixture of romance, surface realism and bits of comic business. Thanks to the strong performances, it just about works.
RCA/Columbia   VHS, Beta   A

Rocky*
US 1976 119m v colour
Sylvester Stallone, Burgess Meredith, Talia Shire, Burt Young, Carl Weathers
D, John G Avildsen

Stallone wrings every drop from his own script about a dim-witted, second-rate boxer who unexpectedly gets a crack at the world title and salvages some self respect. A compelling brew of drama, fairy-tale and fisticuffs.
Warner   VHS, Beta, CED   A

Rocky II
US 1979 117m v colour
Sylvester Stallone, Burgess Meredith, Talia Shire, Carl Weathers, Burt Young
D, Sylvester Stallone

For those who could not have enough of the original *Rocky* here it is more or less all over again, as our ponderous hero has another bruising encounter with the world champ. Needs all Stallone's star quality to keep it afloat.
Warner   VHS, Beta, CED   A

Rocky III
US 1982 99m v colour
Sylvester Stallone, Talia Shire, Burt Young, Burgess Meredith, Carl Weathers
D, Sylvester Stallone

Our hero tries to avenge a humiliating defeat by a brutal slugger and is helped by an old protagonist from the first two *Rocky* films. A rehash of the familiar ingredients, but manages to stay the distance.
Warner   VHS, Beta   A

Romeo and Juliet*

GB/Italy 1968 133m v colour

*Leonard Whiting, Olivia Hussey, John
McEnery, Michael York, Pat Heywood
D, Franco Zeffirelli*

Fresh and colourful rendering of Shakespeare's tragedy of the star-crossed lovers.
It takes big liberties with the text and places
too heavy a burden on its inexperienced
leads – Whiting was aged only 16 and
Hussey 15.
Arena   VHS, Beta   B

La Ronde***

France 1950 93m ● bw

*Anton Walbrook, Simone Signoret, Serge
Reggiani, Danielle Darrieux
D, Max Ophuls*

Arthur Schnitzler's merry-go-round of love,
in which the liaisons between the characters complete a circle. Elegant, stylish, with
lyrical camerawork and many relishable
performances, among which Walbrook's
narrator stands out.
Longman   VHS, Beta   B

The Rose

US 1979 129m ● colour

*Bette Midler, Alan Bates, Frederic
Forrest, Harry Dean Stanton
D, Mark Rydell*

The saga of a rock singer destroyed by drink
and drugs. Loosely derived from the career
of Janis Joplin, it is an excuse to provide a
showcase for the vulgar and brassy talent
of Bette Midler, which it successfully does.
CBS/Fox   VHS, Beta, V2000, Laser   A

Roseland*

US 1977 103m ● colour

*Geraldine Chaplin, Christopher Walken,
Lilia Skala, Teresa Wright
D, James Ivory*

Three stories of romance set in the faded
splendour of the famous New York dance
hall and involving lonely people trying to
come to terms with their memories. The
longest, and best, has Chaplin as an
abandoned wife and Walken as a gigolo.
Virgin   VHS, Beta   A

Ruby Gentry*
US 1952 82m v bw

Jennifer Jones, Charlton Heston, Karl Malden, Josephine Hutchinson
D, King Vidor

Jennifer Jones in one of her most effective "wild women" roles, vowing revenge when the man she loves (Heston) leaves her for a more respectable rival. Steamy Hollywood melodrama, enjoyable mainly for its absurdity.
Guild VHS, Beta A

Ryan's Daughter*
GB 1970 186m v colour

Sarah Miles, Robert Mitchum, John Mills, Trevor Howard, Christopher Jones
D, David Lean

Miles as a romantic Irish girl who marries a dull schoolteacher (Mitchum) twice her age and has an affair with a shell-shocked young officer. Often visually stunning but padded out to enormous length.
MGM VHS, Beta, V2000 A

Salvatore Giuliano**
Italy 1961 123m ● bw

Frank Wolff, Salvo Randone, Federico Zardi
D, Francesco Rosi

The reconstruction through flashbacks, interviews and documentary evidence of the life and death of a notorious Sicilian outlaw. This powerful exposé of violence and corruption sparked off an inquiry into Mafia activities.
Palace VHS, Beta A

Same Time, Next Year
US 1978 113m v colour

Ellen Burstyn, Alan Alda
D, Robert Mulligan

Two lovers carry on an affair over 25 years, meeting once a year in the same Californian motel. Slick adaptation of a Broadway stage hit, studio-bound and episodic, but kept flowing by two excellent performances.
CIC VHS, Beta A

Savages*
US 1972 106m ● colour

Louis Stadlen, Anne Francine, Thayer David, Salome Jens, Neil Fitzgerald
D, James Ivory

Fable about a primitive tribe's encounter with the twentieth century as a group of savages come across a deserted mansion and fall under its civilising influence. An intriguing idea, exploited for less than it is worth.
Palace   VHS, Beta   A

Scarecrow*
US 1973 112m ● colour

Gene Hackman, Al Pacino
D, Jerry Schatzberg

Two of life's losers meet in California and set out to hitch-hike to Pittsburgh with little hope that they will get there. Efficient combination of two favourite 1970s genres, the buddy film and the road movie.
Warner   VHS, Beta   A

Scenes from a Marriage**
Sweden 1973 168m ● colour

Liv Ullman, Erland Josephson, Bibi Andersson, Jan Masjo
D, Ingmar Bergman

A brilliant dissection, full of insight into character and emotion, of an apparently perfect marriage, which starts to founder as the couple realise that they came together out of need rather than love.
Longman   VHS, Beta   B

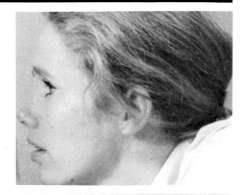

Scrubbers
GB 1982 93m v colour

Amanda York, Chrissie Cotterill, Elizabeth Edmonds, Kate Ingram
D, Mai Zetterling

Would-be realistic examination of life in a girls' borstal is undermined by melodramatic treatment, though the cast – several of whom are non-professional actresses who have spent time in prison – tries to lend conviction.
Thorn EMI   VHS, Beta   A

Scum
GB 1979 96m ● colour

Ray Winstone, Mick Ford, Julian Firth, John Blundell
D, Alan Clarke

Made for the cinema after being banned as a television play, *Scum* is a bleak and violent impression of a boys' borstal with a strong message about the brutalising influence of the institution as grievances build up.
VCL   VHS, Beta   A

Sebastiane*

GB   1976   86m   ● colour

Leonardo Treviglio, Barney James, Neil Kennedy, Richard Warwick
D, Derek Jarman

Low-budget, semi-improvised, independent British feature which explores themes of homosexuality through the story of the martyred soldier of ancient Rome. An original, controversial piece, using Latin dialogue with English subtitles.
Virgin   VHS, Beta   A

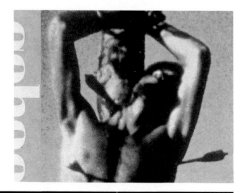

The Servant**

GB   1963   112m   ● bw

Dirk Bogarde, James Fox, Sarah Miles, Wendy Craig
D, Joseph Losey

Penetrating study of upper class decadence in which a spoiled young aristocrat is systematically manipulated and destroyed by his scheming manservant. The teasing script by Harold Pinter cleverly counterpoints Losey's ornate direction.
Thorn EMI   VHS, Beta   C

The Seventh Veil*

GB   1945   94m   v   colour

James Mason, Ann Todd, Herbert Lom, Albert Lieven, Hugh McDermott
D, Compton Bennett

The box-office sensation of its time with Ann Todd as a young concert pianist trying to sort out her emotional problems; Lom is her psychiatrist and Mason her sadistic guardian. Enjoyably daft but it has dated.
Precision   VHS, Beta   C

Shakespeare Wallah*

India   1965   125m   v   bw

Geoffrey Kendal, Laura Liddell, Felicity Kendal, Shashi Kapoor
D, James Ivory

Gentle, atmospheric study of a troupe of English actors bringing Shakespeare to the small towns of India. Slow-going at times but cleverly suggests a country moving away from the old Imperial culture and gaining its own identity.
Palace   VHS, Beta   A

Shoot the Moon*

US   1981   124m   ● colour

Albert Finney, Diane Keaton, Karen Allen, Peter Weller, Dana Hill
D, Alan Parker

Account of the break-up of a marriage and its effect on the children. It might have been more effective if the director had been content to let the perceptive script and fine performances make their own point.
MGM   VHS, Beta, V2000   A

Rocky II

Savages

The Servant

The Shout*
GB 1978 83m ● colour

*Alan Bates, Susannah York, John Hurt,
Robert Stephens
D, Jerzy Skolimowski*

Disturbing, enigmatic piece about the
inmate of a mental asylum who claims to
have acquired the magical aborigine power
to kill by shouting and inflicts his sinister
presence on a philandering composer and
his wife.
Rank   VHS, Beta   A

Le Silence Est d'Or**
France 1947 87m v bw

*Maurice Chevalier, François Périer,
Marcelle Derrien
D, René Clair*

Elegant and witty homage to silent cinema.
It's set in Paris in 1906, with Chevalier
oozing Gallic charm as the middle-aged
film director who falls in love with a young
girl only to lose her to his timid assistant.
Thorn EMI   VHS, Beta   C

Since You Went Away*
US 1944 172m v bw

*Claudette Colbert, Joseph Cotten,
Jennifer Jones, Shirley Temple
D, John Cromwell*

Lush, sentimental and intensely patriotic
study of an American middle-class family
bravely carrying on without father, who is
away fighting in the war. Colbert plays the
mother, with Jones and Temple as the
daughters.
Guild   VHS, Beta   B

The Singer Not the Song
GB 1960 132m v colour

*John Mills, Dirk Bogarde, Mylene
Demongeot, John Bentley
D, Roy Baker*

Bogarde showing his quality with an
authoritative portrayal of a black-clad
Mexican bandit in conflict with Mills'
priest; but what starts out as a promising
study of character becomes too diffuse
under Baker's limp direction.
Rank   VHS, Beta   B

Sky West and Crooked

GB   1965   101m   v   colour

*Hayley Mills, Ian McShane, Laurence
Naismith, Geoffrey Bayldon*
D, John Mills

Hayley Mills, directed by father John, as a
mentally retarded 17-year-old girl who is
rescued from a river by a young gypsy and
falls in love with him. Nicely played
romantic tear jerker with darker under-
tones.
Rank   VHS, Beta   A

Sophie's Choice*

US   1982   140m   v   colour

*Meryl Streep, Kevin Kline, Peter
MacNicol, Rita Karin, Stephen D
Newman*
D, Alan J Pakula

A Polish girl living in New York carries a
guilty secret from her wartime experience
in a concentration camp into her relationships with two men. Oscar-winning performance from Meryl Streep.
Precision   All systems   B

Stage Struck

US   1957   91m   v   colour

*Henry Fonda, Susan Strasberg, Herbert
Marshall, Joan Greenwood*
D, Sidney Lumet

Re-make of an old Katharine Hepburn
vehicle, *Morning Glory*, about a young
actress from the sticks who is determined
to make it on Broadway. Strasberg, unfortunately, is no Hepburn. With *The Girl Most
Likely*, a 1957 musical.
Kingston   VHS, Beta   A

Stardust

GB   1974   108m   v   colour

*David Essex, Adam Faith, Larry Hagman,
Marty Wilde*
D, Michael Apted

A disenchanted look at the rise and fall of a
1960s pop singer and his group, based
loosely on the story of the Beatles but
exuding little of their charm and personality. Essex as the doomed hero; Faith his
cynical road manager.
Thorn EMI   VHS, Beta   A

Steelyard Blues

US   1972   89m   v   colour

*Donald Sutherland, Jane Fonda, Peter
Boyle, Howard Hesseman*
D, Alan Myerson

Sutherland is an ex-convict who steals cars
to race in demolition derbies; Fonda is his
call-girl friend and Boyle his district attorney brother, who tries to take him in hand.
Quirky, often funny, but as aimless as its
hero.
Warner   VHS, Beta   A

Stevie*
US/GB   1978   102m   v   colour

Glenda Jackson, Mona Washbourne,
Trevor Howard, Alec McCowen
D, Robert Enders

The uneventful life in a north London terrace house of the novelist and poetess Stevie Smith, based on the play by Hugh Whitemore and confined virtually to one room. The acting carries it, especially Washbourne as the kindly aunt.
Home Video Productions   VHS, Beta   B

Straight Time
US   1978   114m   ●   colour

Dustin Hoffman, Theresa Russell, Gary
Busey, Harry Dean Stanton
D, Ulu Grosbard

A slackly handled portrait of a born loser, a psychotic long term offender who is paroled after six years in jail but quickly abuses his new-found freedom. Hoffman tries hard to give him substance.
Warner   VHS, Beta   A

Straw Dogs
GB   1971   113m   ●   colour

Dustin Hoffman, Susan George, Peter
Vaughan, David Warner
D, Sam Peckinpah

An American postgraduate student and his wife move into a quiet Cornish village but find it less than idyllic. An explicit rape and other violent scenes led the film to be banned by several British local authorities.
Guild   VHS, Beta, V2000   A

Stromboli*
Italy   1949   79m   v   bw

Ingrid Bergman, Mario Vitale, Renzo
Cesana
D, Roberto Rossellini

Bergman as a Lithuanian girl made homeless by the war. She marries an Italian fisherman but finds difficulty adjusting to life on his primitive island; the volcano makes a brooding backdrop. With *While the City Sleeps* (qv).
Kingston   VHS, Beta   A

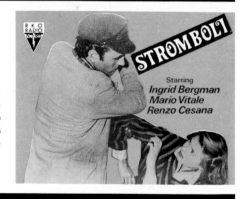

Stroszek*
W Germany   1977   107m   v   col

Bruno S, Eva Mattes, Clemens Scheitz
D, Werner Herzog

Bruno S (discovered by Herzog while working as a lavatory attendant) gives a remarkable performance as a simple-minded jailbird who abandons Berlin with his prostitute girlfriend to start a new life in Wisconsin.
Palace   VHS, Beta   A

171

Stardust

Stevie

The Stuntman

The Stunt Man*
US   1979   129m   ● colour

Peter O'Toole, Steve Railsbeck, Barbara Hershey, Allen Goorwitz
D, Richard Rush

Unusual, compelling, ambiguous piece about a Vietnam veteran on the run from the police who accidentally causes the death of a stunt man on a film and is hired in his place. O'Toole in fine form as the manipulating director.
Guild   VHS, Beta, V2000   A

Summer of '42*
US   1971   98m   ● colour

Jennifer O'Neill, Gary Grimes, Jerry Houser, Oliver Conant
D, Robert Mulligan

A 15-year-old boy's sexual initiation with a young war bride on a holiday island off New England in 1942. Sentimental and nostalgic, told in flashback, with misty camerawork and a lush score from Michel Legrand.
Warner   VHS, Beta   A

Summer with Monika*
Sweden   1952   98m   ● colour

Harriet Andersson, Lars Ekborg, John Harryson
D, Ingmar Bergman

Efficient, but minor Bergman work. A self-centred shopgirl and a young workman run away to spend an island holiday; she becomes pregnant, tires of the relationship and eventually abandons him with the baby.
Longman   VHS, Beta   B

Table For Five
US   1983   123m   v   colour

Jon Voight, Richard Crenna, Marie Christine Barrault
D, Robert Lierberman

Unashamed tearjerker. A divorced father of three tries to "make a comeback" into his children's lives by taking them on a Mediterranean cruise, but finds that they have changed since he last saw them four years before.
CBS/Fox   VHS, Beta   A

A Tale of Two Cities
GB   1958   112m   □   bw

Dirk Bogarde, Dorothy Tutin, Christopher Lee, Rosalie Crutchley
D, Ralph Thomas

Solid, uninspired adaptation of Charles Dickens' novel of the French Revolution. A characteristically intelligent performance from Bogarde as the young English lawyer, Sydney Carton, going to the guillotine in another man's place.
Rank   VHS, Beta   A

Taxi Driver**
US 1976 116m ● colour
*Robert de Niro, Jodie Foster, Cybill
Shepherd, Paul Boyle
D, Martin Scorsese*

De Niro as an embittered Vietnam war
veteran turned urban guerilla, with Jodie
Foster the 12-year-old prostitute he tries to
save. The savage intensity is not easy to
like, but there is much to admire in this
production.
RCA/Columbia   VHS, Beta, CED   R

The Tempest
US 1982 136m v colour
*John Cassavetes, Gena Rowlands, Susan
Sarandon, Vittorio Gassman
D, Paul Mazursky*

Interesting, but only patchily successful,
attempt to work the themes of Shakespeare's play into a modern story of a
Manhattan architect who flees his marriage for a lonely Greek island and the
storm that brings reconciliation.
RCA/Columbia   VHS, Beta   R

The Tempest*
GB 1979 95m v colour
*Toyah Willcox, Heathcote Williams, Karl
Johnson, Peter Bull
D, Derek Jarman*

Artful, inventive, highly decorated rendering of Shakespeare's late play. Atmo-
spherically set in a crumbling mansion, it
combines a faithful adherence to the
original text with images of modern pop
culture.
Palace   VHS, Beta   A

Tess**
France/GB 1979 164m v col
*Nastassia Kinski, Leigh Lawson, Peter
Firth, John Collin
D, Roman Polanski*

Thomas Hardy's tragic story of the peasant
girl destroyed by her faithless suitors
comes to the screen with a clean narrative
flow. Normandy stands in for nineteenth
century Dorset and there is a touching
central performance.
Thorn EMI   VHS, Beta   A

That'll Be the Day
GB 1973 86m v colour
*David Essex, Ringo Starr, Rosemary
Leach, James Booth, Billy Fury
D, Claude Whatham*

Downbeat look at the British youth scene of
the 1950s. Essex makes a creditable
appearance in his first film part as a
teenage drifter who deserts his wife and
child to become a pop star.
Thorn EMI   VHS, Beta   A

They Shoot Horses Don't They?*

US 1969 108m v colour

Gig Young, Jane Fonda, Susannah York, Michael Sarrazin, Red Buttons
D, Sydney Pollack

The hopes and the despair of contestants in a Hollywood dance marathon in the 1930s. Pretentious direction is offset by a superb, Oscar-winning performance from Gig Young as the cynical master of ceremonies.
Rank   VHS, Beta   B

36 Chowringhee Lane**

India   1981   122m   v   colour

Jennifer Kendal, Debashree Roy, Dhritiman Chatterjee, Geoffrey Kendal
D, Aparna Sen

Poignant study of a lonely Anglo-Indian schoolteacher in Calcutta, living on her memories and realising that her time is past. A distinguished directorial debut by the actress, Aparna Sen, rich in perceptive detail.
Monte Video   VHS, Beta   A

This Sporting Life***

GB   1963   129m   ●   bw

Richard Harris, Rachel Roberts, Alan Badel, Colin Blakely
D, Lindsay Anderson

One of the British cinema's supreme achievements, charting the rise and decline of a rugby league footballer and his doomed affair with a miner's widow; the emotional tensions are reflected in the bleak northern landscapes.
Rank   VHS, Beta   A

The Three Sisters*

GB   1970   158m   v   colour

Joan Plowright, Jeanne Watts, Louise Purnell, Alan Bates, Derek Jacobi
D, Laurence Olivier

Straightforward adaptation by Olivier of his stage production of Chekhov's play. Three sisters in a Russian provincial town at the turn of the century are bored by their life and dream of returning to Moscow.
Thorn EMI   VHS, Beta   A

Thunderbolt and Lightfoot*

US   1974   115m   ●   colour

Clint Eastwood, Jeff Bridges, George Kennedy, Catherine Bach
D, Michael Cimino

Violent tale of a bank robber on the run (Eastwood) who strikes up a friendship with a young drifter (Bridges) on his way to recover the loot, only to discover that the building where he hid it is no longer there.
Warner   VHS, Beta   A

Tiara Tahiti

GB   1962   98m   v   colour

John Mills, James Mason, Herbert Lom, Claude Dauphin
D, William T Kotcheff

Former army colonel building a hotel in Tahiti runs into trouble from an old enemy, a captain he had cashiered for smuggling. Amiable knockabout, with flashes of humour, and held together by its two stars.
Rank   VHS, Beta   A

Taxi Driver

They Shoot Horses Don't They?

Tiara Tahiti

Tiger Bay*
GB 1959 105m v bw

Hayley Mills, John Mills, Horst Buchholz, Megs Jenkins
D, J Lee-Thompson

Memorable screen debut by the 12-year-old Hayley Mills as a waif in the Cardiff docks who watches a Polish seaman (Buchholz) killing his girlfriend and tries to protect him against the police (led by father, John Mills).
Rank   VHS, Beta   A

To Kill a Mockingbird*
US 1962 129m v bw

Gregory Peck, Mary Badham, Philip Alford, John Megna, Frank Overton
D, Robert Mulligan

Quietly effective indictment of race prejudice in the American south during the Depression, with Peck as the lawyer defending a negro accused of assaulting and raping a white girl. Notable for the unforced playing of the child actors.
RCA/Columbia   VHS, Beta   A

Tom Jones*
GB 1963 122m ● colour

Albert Finney, Susannah York, Hugh Griffith, Edith Evans
D, Tony Richardson

Enjoyable-in-parts romp from Henry Fielding's novel about the amorous adventures of a squire's adopted son in eighteenth century England. Some fruity performances but the director's cinematic trickery becomes tiresome.
Thorn EMI   VHS, Beta   A

True Confessions
US 1981 m v colour

Robert de Niro, Robert Duvall, Charles Durning, Kenneth McMillan
D, Ulu Grosbard

The involvement of two brothers, a detective (Duvall) and a Catholic priest (De Niro), in a Los Angeles murder case in the 1940s. Effectively acted but weakly directed and the moral ambiguities could have been more fully explored.
CBS/Fox   VHS, Beta   A

The Turning Point
US 1977 114m v colour

Anne Bancroft, Shirley Maclaine, Mikhail Baryshnikov, Leslie Browne
D, Herbert Ross

Superior soap opera, with two commanding central performances. The ageing star of the American Ballet Theater (Bancroft) renews acquaintance with an old friend (Maclaine) whose daughter wants to be a dancer.
CBS/Fox   VHS, Beta, V2000   A

An Unmarried Woman*
US 1977 119m ● colour

Jill Clayburgh, Alan Bates, Michael Murphy, Cliff Gorman, Pat Quinn
D, Paul Mazursky

Vibrant acting by Clayburgh as a woman who learns to cope after being abandoned by her husband. Hailed in some quarters as a modern feminist tract but really a throwback to old-fashioned Hollywood romance.
CBS/Fox   VHS, Beta, Laser   A

Urban Cowboy
US 1980 125m v colour

John Travolta, Debra Winger, Scott Glenn, Madolyn Smith
D, James Bridges

Travolta doing his disco thing but otherwise finding it hard to register as the ordinary kid from the Texas sticks trying to make it in the big city. Nice performance from Winger as the girl he briefly marries.
CIC   VHS, Beta   A

Valley of the Dolls
US 1967 118m ● colour

Barbara Parkins, Patty Duke, Susan Hayward, Sharon Tate, Paul Burke
D, Mark Robson

Overheated soap opera from the Jacqueline Susann bestseller about three young innocents plunging into the nasty world of showbusiness and getting mauled. Glossily presented but emotionally empty.
CBS/Fox   VHS, Beta   A

The Verdict
US 1982 128m v colour

Paul Newman, James Mason, Charlotte Rampling, Jack Warden, Milo O'Shea
D, Sidney Lumet

Solid, satisfying courtroom drama sustained by Lumet's careful direction and the professionalism of the protagonists: Newman as drunken Boston attorney trying to salvage his career faces Mason as the wily defence lawyer.
CBS/Fox   VHS, Beta   A

Victim*
GB   1961   96m   ● bw

Dirk Bogarde, Sylvia Sims, Dennis Price,
Nigel Stock, Peter McEnery
D, Basil Dearden

Courageous (for its time) plea for homosexual law reform with Bogarde as a
barrister prepared to risk his marriage and
career to expose a blackmailer after one of
his former lovers commits suicide.
Rank   VHS, Beta   B

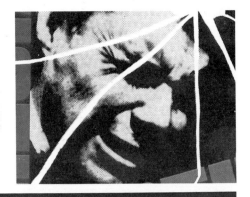

The Virgin Spring**
Sweden   1959   88m   ● bw

Max von Sydow, Birgitta Pettersson,
Birgitta Valberg, Gunnel Lindblom
D, Ingmar Bergman

Starkly beautiful rendering of a medieval
legend of the young virgin whose rape and
murder by three brigands is brutally
avenged by her father; and of the spring
which suddenly gushes forth from the spot
where she died.
Longman   VHS, Beta   B

Watership Down*
GB   1978   88m   □ colour

Voices: John Hurt, Richard Briers, Ralph
Richardson, Zero Mostel
D, Martin Rosen

Cartoon version of Richard Adams' best-
seller about a group of rabbits seeking a
new and safer home after the destruction
of their warren. Carefully and faithfully
done, though the well-known voices can be
a distraction.
Thorn EMI   VHS, Beta   A

Week-end**
France   1967   95m   v colour

Mireille Darc, Jean Yanne, Jean-Pierre
Kalfon, Valérie Lagrange
D, Jean-Luc Godard

Godard's biting, surreal look at modern
capitalist society. A middle-class couple on
a visit to mother-in-law pick their way
through blazing cars and bloody corpses
and are captured by cannibalistic hippies.
Capstan   VHS, Beta   A

Whistle Down the Wind**

GB 1961 95m v bw

Hayley Mills, Bernard Lee, Alan Bates,
Diane Holgate
D, Bryan Forbes

Bates as a murderer on the run who is
sheltered by three children who believe
him to be Jesus. A touching study of
childhood innocence, set against a brooding Northern landscape and with perfectly
judged performances.
Rank VHS, Beta A

White Dog*

US 1981 90m v colour

Kristy McNichol, Paul Winfield, Burl Ives,
Jameson Parker
D, Samuel Fuller

Sharp parable on racialism. A handsome
Alsatian has been trained to attack black
people and has the humans searching their
consciences as they try to decide what to
do.
CIC VHS, Beta A

Who'll Stop the Rain?

US 1978 125m ● colour

Michael Moriarty, Tuesday Weld, Nick
Nolte, Anthony Zerbe
D, Karel Reisz

A newspaperman makes his protest
against the Vietnam war by shipping
heroin back to the United States through a
navy friend, but thugs are soon on their
trail. Efficiently mounted piece, not very
sure of its intentions.
Warner VHS, Beta A

Who's Afraid of Virginia Woolf?**

US 1966 124m ● bw

Richard Burton, Elizabeth Taylor, George
Segal, Sandy Dennis
D, Mike Nichols

Compulsive screen adaptation of Edward
Albee's play about a warring college
professor and his wife who inflict their
emotional agony on two young guests.
Taylor and Burton in fine bitchy form.
Warner VHS, Beta A

Whose Life is it Anyway?

US 1981 115m ● colour

Richard Dreyfuss, John Cassavetes,
Christine Lahti, Bob Balaban
D, John Badham

Brian Clark's unexpected stage hit about a
young sculptor paralysed after a car crash
who pleads to be able to die in dignity. Not
really a big screen subject, though Dreyfuss does his best.
MGM/UA VHS, Beta V2000, Laser A

The Wild Heart
GB 1950 88m v colour

Jennifer Jones, David Farrar, Cyril
Cusack, Sybil Thorndike
D, Michael Powell/Emeric Pressburger

Jennifer Jones was imported from Hollywood to play a Shropshire girl courted by the local squire in Victorian England and made a fair stab at it. None too successful attempt to adapt a Mary Webb novel. Also titled *Gone to Earth*.
Guild VHS, Beta A

The Wild Party
US 1974 90m ● colour

James Coco, Raquel Welch, Perry King,
Tiffany Bolling
D, James Ivory

Silent film star on the slide throws a lavish Hollywood party to promote his new picture but the occasion becomes a drunken orgy. Muddled and uneven look at the film colony; Raquel Welch superb as the star's mistress.
Rank VHS, Beta B

The Turning Point

Yol

181

Wild Strawberries***

Sweden 1957 89m v bw

Victor Sjöström, Bibi Andersson, Ingrid Thulin, Gunnar Björnstrand, Max von Sydow

D, Ingmar Bergman

This is ostensibly an account of a car journey taken by a 78-year-old professor, Isak Borg, from his home to the university to receive an honorary doctorate. It is a journey both into the future and into the past, allowing the old man to look back over his life and reveal his character and personality. A dream sequence at the start of the film, in which Borg sees his own body in a coffin which has fallen from a hearse, establishes that he is close to death; his pilgrimage to the university may be a last chance to take stock. Another dream, set in the strawberry field which gives the film its title, recalls his first love,

the girl who rejected him and married his brother. Gradually his past is stripped away, revealing a lonely and selfish individual, whose professional standing as marked by the award of the doctorate is in sad contrast to the failures of his personal relationships. Bergman is a director distinguished by the bleakness of his vision and *Wild Strawberries* does little to contradict this. But while it may be hard to like Bergman's films, it is impossible not to admire them; and *Wild Strawberries* is among his finest achievements, a perfectly controlled piece of cinema in which every frame makes its point and acted with a faultless precision. Borg is played by Victor Sjöström, a distinguished Swedish director and actor with a career going back to the early silent era: *Wild Strawberries* was his last film and could not have made a more moving farewell.

Longman   VHS, Beta   B

The Winslow Boy*
GB 1948 112m v bw

*Robert Donat, Cedric Hardwicke,
Margaret Leighton, Frank Lawton
D, Anthony Asquith*

Carefully crafted and absorbing adaptation of Terence Rattigan's play about a boy naval cadet expelled for stealing a postal order and the fight to clear him; with a fine performance from Donat as the defending barrister.
Thorn EMI VHS, Beta A

Women in Love*
GB 1969 130m ● colour

*Glenda Jackson, Jennie Linden, Alan Bates, Oliver Reed
D, Ken Russell*

D H Lawrence's dense novel of sexual awakening in a Midlands coal town is anchored firmly in period and setting. Russell for once showed the virtues of restraint and Jackson won an Oscar as the tragic Gudrun.
Warner VHS, Beta A

Woyzeck**
WG 1979 80m v colour

*Klaus Kinski, Eva Mattes, Wolfgang Reichmann, Willy Semmelrogge
D, Werner Herzog*

The tragedy of the simple soldier who is humiliated by everyone around him and is finally driven to murder; fashioned into a brilliant, compelling piece of cinema. (With *How Much Wood Would a Woodchuck Chuck*.)
Palace VHS, Beta A

The Year of Living Dangerously*
Australia 1982 111m v colour

*Mel Gibson, Sigourney Weaver, Linda Hunt, Michael Murphy, Bill Kerr
D, Peter Weir*

Gibson as a greenhorn Australian reporter in Jakarta just before the coup that toppled Sukarno in 1965, manoeuvred into an affair and facing the dilemma of whether to betray confidences for a story.
MGM VHS, Beta A

Yol**
Switz/WG 1982 105m ● col

*Tarik Akan, Halin Ergün, Mecmettin Cobanoglü, Serif Sezer
D, Serif Gören*

The tragic study of a group of prisoners on temporary release who find that their freedom is an illusion; remarkable not only for its filmic qualities but for the fact that it was masterminded from jail by the Turkish director, Yilmaz Güney.
Polygram VHS, Beta A

Zabriskie Point
US 1969 108m ● colour

*Mark Frechette, Daria Halprin, Rod Taylor, Paul Fix
D, Michelangelo Antonioni*

A sumptuously photographed but dramatically empty look at the American youth protest of the 1960s by a once formidable Italian director running sadly out of creative steam. The unknown leads are little help.
MGM VHS, Beta B

HALLOWEEN II

From the people who brought you HALLOWEEN...
more of the night HE came home

An American Werewolf in London*

GB   1981   93m   ● colour

David Naughton, Jenny Agutter, Griffin Dunne, John Woodvine
D, John Landis

A werewolf rampages through Piccadilly Circus and other bits of tourist Britain. Writer/director Landis shows a happy knack of mixing horror with comedy and pathos; excellent special effects.
Polygram   All systems   A

The Amityville Horror

US   1979   118m   ● colour

James Brolin, Margot Kidder, Rod Steiger, Don Stroud
D, Stuart Rosenberg

Newly-weds Brolin and Kidder buy a haunted house in Amityville and soon wish they hadn't. Nothing original in the subject-matter but director Rosenberg's treatment is fetchingly spirited. Also available: *Amityville II – The Possession*.
Guild   VHS, Beta, V2000, CED   A

And Now the Screaming Starts!

GB   1973   87m   ● colour

Peter Cushing, Stephanie Beacham, Herbert Lom, Patrick Magee
D, Roy Ward Baker

Murderous fun and games in an ancestral home with a turbulent past. Stephanie Beacham is the screaming heroine, moving into the house as a new bride, while Cushing provides his usual touch of style.
Guild   VHS, Beta, V2000   A

Asylum

GB   1972   95m   ● colour

Patrick Magee, Robert Powell, Geoffrey Bayldon, Barbara Perkins, Sylvia Sims
D, Roy Ward Baker

Four tales about voodoo, schizophrenia, deadly mannikins and a tailor's dummy that comes to life, from stories by *Psycho* author Robert Bloch. Uneven, but the asylum framework is clever and the best scenes suitably outrageous.
Guild   VHS, Beta   A

The Awakening
GB 1980 100m ● colour

Charlton Heston, Susannah York, Jill
Townsend, Stephanie Zimbalist
D, Mike Newell

Public response to the Tutankhamun exhi-
bition supposedly inspired this piece of
British mummified nonsense, derived from
Bram Stoker. The cast seems burdened by
the absurdities; pleasingly photographed
Egyptian locations.
Thorn EMI VHS, Beta A

Beast in the Cellar
GB 1970 90m ● colour

Flora Robson, Beryl Reid, Tessa Wyatt,
John Hamill, T P McKenna
D, James Kelly

Robson and Reid as two sweet old Lancashire spinsters who keep a guilty secret
bricked up in their cellar. Writer/director
Kelly spends too much time twiddling his
thumbs but manages a satisfyingly grisly
climax.
Guild VHS, Beta A

The Birds**
US 1963 119m ● colour

Rod Taylor, Tippi Hedren, Jessica Tandy,
Suzanne Pleshette, Ethel Griffies
D, Alfred Hitchcock

Our feathered friends suddenly turn
enemies, ferociously attacking an isolated
Californian community. The human drama
takes second place to a mounting aura of
fear and there are some memorable special
effects.
CIC VHS, Beta A

Blood from the Mummy's Tomb*
GB 1971 90m ● colour

Andrew Keir, Valerie Leon, James
Villiers, Hugh Burden, George Coulouris
D, Seth Holt

An archaeological expedition bears hideous,
supernatural fruit; fascinating attempt to
breathe life into a well-used genre, based
on the same Bram Stoker novel that
prompted *The Awakening* (qv).
Thorn EMI VHS, Beta A

Blood on Satan's Claw

GB   1970   93m   ● colour

Patrick Wymark, Linda Hayden, Barry
Andrews, Avice Landon, Simon Williams
D, Piers Haggard

Macabre goings-on in a seventeenth century English village where the children are
possessed by the devil. Tails off after a
promising start but the climax is worth
waiting for and there are other moments to
cherish.
Guild   VHS, Beta   A

The Brood

Canada   1979   91m   ● colour

Oliver Reed, Samantha Eggar, Art
Hindle, Cindy Hinds
D, David Cronenberg

Only Cronenberg could invent the science
of "psychoplasmics" and only he could
come up with such a tale about mental
derangement and killer children. A gruesome piece, presented with skill by a
provocative talent.
Alpha   VHS, Beta, V2000   A

A Candle for the Devil

Spain   1973   82m   ● colour

Judy Geeson, Aurora Bautista,
Esperanza Roy, Vic Winner
D, Eugene Martin

A sexually frustrated hotel proprietor murders her guests in the interests of godli-
ness. All the old clichés are brought out and
lovingly polished; the acting and the
Spanish location photography are high
class.
Vampix   VHS, Beta, V2000   A

Carrie*

US   1976   98m   ● colour

Sissy Spacek, Piper Laurie, Amy Irving,
William Katt, John Travolta
D, Brian de Palma

Energetic frightener from Stephen King's
novel about a repressed schoolgirl with
telekinetic powers who takes macabre
revenge on her taunting colleagues. Stylish
nonsense, with echoes of Hitchcock.
Warner   VHS, Beta, V2000, CED   A

Cat People

US   1982   114m   ● colour

Nastassia Kinski, Malcolm McDowell,
John Heard, Annette O'Toole, Ruby Dee
D, Paul Schrader

Kinski and McDowell play a brother and
sister with a weird feline ancestry in an
interesting attempt to rework the 1942
classic; strong New Orleans atmosphere
and explicit eroticism.
CIC   VHS, Beta, V2000, Laser   A

Cauldron of Blood

Spain/US   1968   87m   ● colour

Boris Karloff, Viveca Lindfors, Jean-
Pierre Aumont, Rosenda Monteros
D, Edward Mann

Karloff, nearing the end of his career, is a
blind sculptor who draws inspiration from
human remains loyally acquired by his
sadistic wife (Lindfors). Low-budget treatment of a clever idea, weighed down by
symbolism.
Vampix   VHS, Beta, V2000   A

The Changeling

Canada   1979   113m   ● colour

George C Scott, Melvyn Douglas, Trish
van Devere, John Colicos
D, Peter Medak

Widowed composer Scott buys a grand
Seattle house with bad vibrations caused
by a murdered child. Fair ghost, flashily
directed, and containing one of the last
performances of Melvyn Douglas, as a
senator with a shady past.
VTC   VHS, Beta   A

Circus of Horrors

GB   1960   87m   ● colour

Anton Diffring, Erika Remberg, Yvonne
Monlaur, Donald Pleasence, Jane Hylton
D, Sidney Hayers

Anton Diffring cultivates his villainy as a
plastic surgeon up to no good in a circus
stocked with criminals; he is finally undone
when one of his victims catches up with
him. Unsubtle stuff, robustly handled.
Thorn EMI   VHS, Beta   A

City of the Dead

GB   1960   78m   ● bw

Patricia Jessel, Betta St John, Dennis
Lotis, Christopher Lee, Valentine Dyall
D, John Moxey

A student of the occult bites off more than
she can chew as she researches witchcraft
in a crumbling Massachusetts village.
Low-budget piece, with sufficient spirit and
atmosphere to compensate for clodhop-
ping moments.
Intervision   VHS, Beta   A

Curse of the Crimson Altar
GB 1968 89m ● colour

Boris Karloff, Christopher Lee, Rupert Davies, Mark Eden, Barbara Steele
D, Vernon Sewell

Eden searching for a missing brother: the trail is strewn with diabolic rituals, a tame wild party and horror queen Steele dressed to kill in green make-up and a ram headpiece. Karloff's last film.
Vampix VHS, Beta, V2000 A

Damien: Omen Two
US 1978 103m ● colour

William Holden, Lee Grant, Jonathan Scott-Taylor, Robert Foxworth
D, Don Taylor

First of two sequels to *The Omen* (qv) with the anti-Christ, now a teenager, enrolled at a military academy. Otherwise the mixture as before: a story of devilment and mounting suspicions, told with grisly aplomb.
CBS/Fox VHS, Beta, V2000, Laser A

Death Line*
GB 1972 84m ● colour

Donald Pleasence, Christopher Lee, David Ladd, Sharon Gurney
D, Gary Sherman

A poor advertisement for London Transport as an inspector (Pleasence) investigates a series of disappearances and murders in the bowels of the underground system. Unusually inventive horror, relieved by a sense of humour.
Rank VHS, Beta A

Dr Jekyll and Mr Hyde**
US 1920 63m ● bw

John Barrymore, Nita Naldi
D, John S Robertson

Fascinating silent version of Robert Louis Stevenson's classic, with John Barrymore (then at the peak of his stage career) giving a bravura performance as the doctor. The transformation scenes use few special effects.
Polygram VHS, Beta B

Dr Jekyll and Sister Hyde
GB 1971 94m ● colour

Ralph Bates, Martine Beswick, Gerald Sim, Lewis Fiander, Dorothy Alison
D, Roy Ward Baker

This old story gets a novel sex-change twist – the result of Jekyll taking female hormones. Martine Beswick takes care of Sister; Ralph Bates makes a persuasive Doctor. The production is conventional but assured.
Thorn EMI VHS, Beta A

Dracula*
US 1931 73m ● bw

Bela Lugosi, Helen Chandler, David Manners, Dwight Frye
D, Tod Browning

Later adventures of the Transylvanian count featured better bats, livelier direction and less theatrical ham; here the stage source looms too large. Essential viewing for moments of eerie splendour and the sepulchral Lugosi.
CIC VHS, Beta B

Dracula
US 1979 109m ● colour

Frank Langella, Laurence Olivier, Donald Pleasence, Kate Nelligan, Trevor Eve
D, John Badham

Stylish and romanticised version of the old horror stand-by, which has the merit of not asking you to take it too seriously. Frank Langella is an unusually debonair Dracula, Olivier an unsubtle Van Helsing.
CIC VHS, Beta, Laser A

The Entity
US 1981 119m ● colour

Barbara Hershey, Ron Silver, David Labiosa, George Coe, Jacqueline Brookes
D, Sidney J Furie

An invisible thing assaults Hershey. Psychiatrist Silver thinks it is a figment of her imagination – more likely a figment of film makers desperate to shock. Strong meat, skilfully served.
CBS/Fox VHS, Beta A

Eraserhead**
US 1976 100m ● bw

John Nance, Charlotte Stewart, Allen Joseph, Jeanne Bates
D, David Lynch

Extraordinary film by the future director of *The Elephant Man* (qv). A journey through a nightmare world of mutating forms, bizarre phenomena and raging urban angst. Grotesquely funny, shocking and technically excellent.
Palace VHS, Beta A

The Exorcist*
US 1973 115m ● colour

Ellen Burstyn, Max Von Sydow, Jason Miller, Linda Blair, Lee J Cobb
D, William Friedkin

The Devil makes a spectacular come-back in this very tall story about a possessed girl (Linda Blair) with an appalling urge to kill. A showy piece of work but undeniably effective. Also available: *Exorcist II – The Heretic.*
Warner VHS, Beta A

Eyes Without a Face**
France/Italy 1959 90m ● bw

Pierre Brasseur, Alida Valli, Edith Scob, François Guérin
D, Georges Franju

A surgeon tries repairing his mutilated daughter with other people's faces. Franju's acute sense of pulp poetry and the horribly beautiful transforms what might have been merely unpleasant into a haunting minor classic.
Thorn EMI VHS, Beta C

The Fall of the House of Usher*
US 1960 78m ● colour
Vincent Price, Myrna Fahey, Mark Damon, Harry Ellerbe
D, Roger Corman
Poe's story about a tormented household is brought to cinematic life with low-budget enterprise and imagination. Vincent Price excels as the necrophiliac brother who entombs his sister and pays the penalty.
Guild VHS, Beta A

The Fog*
US 1979 91m ● colour
Adrienne Barbeau, Hal Holbrook, John Houseman, Janet Leigh
D, John Carpenter
A fog full of seafaring ghosts wraps itself around a small town on the Californian coast. Director Carpenter (*Halloween*, qv) dawdles too much but creates a dark, clammy atmosphere and delivers some sharp shocks.
Embassy All systems A

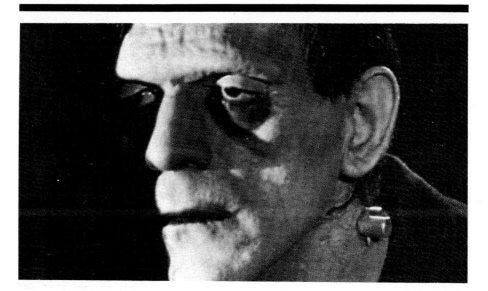

Frankenstein***
US 1931 71m ● bw
Boris Karloff, Colin Clive, Mae Clarke, John Boles, Edward Van Sloan
D, James Whale
Frankenstein has left indelible marks on cinema history. It was the film (with *Dracula*, which appeared in the same year) that launched Universal Studios on a famous horror cycle which lasted through the 1930s and included *The Mummy, The Invisible Man* as well as two Frankenstein sequels. It was the first horror film for the English-born William Pratt, better known as Boris Karloff; he went on to make nearly 50. It was the first of many pictures — some of them straight adaptations, most not — inspired by Mary Shelley's novel which had been published in 1818. A common mistake is to confuse Frankenstein, the mad doctor, with the monster he creates out of the organs of dead bodies. In the film Frankenstein is played by Colin Clive, while Karloff's monster represents a considerable feat by the make-up artist, Jack Pierce. He spent four hours a day applying his magic touches, including a huge box-like rubber head held together by metal clamps; and at the end of it all, Karloff already a six-footer, had "grown" by some 18 inches. *Frankenstein* may lack in technical sophistication compared with today's films, but it is still a marvellous movie in its own right. It has two great virtues which contemporary film makers could well heed: it tells its story with the utmost economy and it manages to convey horror effectively without actually showing it. The director, James Whale, was also an Englishman; he came to Hollywood to make a film version of *Journey's End* and he stayed to add a notable contribution to the horror genre.
CIC VHS, Beta A

191

Frogs*

US 1972 91m ● colour

Ray Milland, Joan Van Ark, Sam Elliott,
Adam Roarke, Judy Pace
D, George McCowan

Nature rebels on Ray Milland's man-made island in the American South. An accomplished and ingenious romp with a low special effects bill since the monsters are all frogs, toads and lizards, caught on the hop.
Guild VHS, Beta, V2000 A

The Fury

US 1978 118m ● colour

Kirk Douglas, John Cassavetes, Carrie
Snodgress, Charles Durning
D, Brian de Palma

A startling, bloody and complex tale of abused psychic powers and political ambitions, told by a flashy director. Kirk Douglas supplies a touch of class as the government boffin searching for his missing son.
CBS/Fox VHS, Beta, V2000, Laser A

Halloween*

US 1978 91m ● colour

Donald Pleasence, Jamie Lee Curtis,
Nancy Loomis, P J Soles
D, John Carpenter

Murders galore in a small Illinois town, staged with skilful precision by a director who knows how to scare. Jamie Lee Curtis, makes an admirable schoolgirl heroine; Pleasence in his usual chilly form as the local doctor.
Videoform VHS, Beta, V2000 A

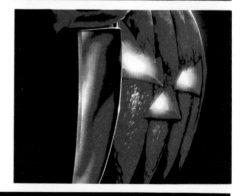

Halloween II

US 1981 88m ● colour

Donald Pleasence, Jamie Lee Curtis,
Charles Cyphers, Jeffrey Kramer
D, Rick Rosenthal

Less sophisticated sequel, with the psychopathic killer continuing where he left off on the same Halloween night. The camera spends too much time wandering through corridors but the basic formula pulls the film through.
Thorn EMI VHS, Beta A

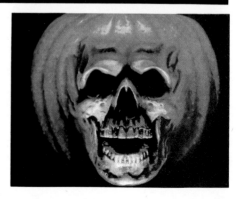

Halloween III: Season of the Witch

US   1983   91m   ●   colour

Tom Atkins, Stacey Nelkin, Dan O'Herlihy, Ralph Strait, Michael Currie
D, Tommy Lee Wallace

New maniac, victims and location, but the recipe remains unchanged: ingenious, bloody, night-time killings in a small town. O'Herlihy as a toy factory owner with fiendish plans for murder.
Thorn EMI   VHS, Beta   A

Hands of the Ripper

GB   1971   85m   ●   colour

Angharad Rees, Eric Porter, Dora Bryan, Jane Merrow, Derek Godfrey
D, Peter Sasdy

Jack the Ripper's daughter (Rees) grows up with her father's murderous characteristics. There are stabbings, impalings and leaps from on high, though the film is best at quiet moments of foreboding.
Rank   VHS, Beta, Laser   A

House of Wax*

US   1953   85m   ●   colour

Vincent Price, Carolyn Jones, Paul Picerni, Phillis Kirk, Frank Lovejoy
D, André de Toth

Crude but vigorous re-make of a 1930s classic, *Mystery of the Wax Museum*; originally released in 3-D. Vincent Price relishes his first big horror part as the crazed wax sculptor with novel working methods.
Warner   VHS, Beta   A

The Hunchback of Notre Dame*

US   1923   93m   ●   bw

Lon Chaney, Patsy Ruth Miller, Norman Kerry, Ernest Torrence, Gladys Brockwell
D, Wallace Worsley

Spectacular silent production of Victor Hugo's novel about the deformed bell-ringer, though without Chaney it would be heavy going. The extravagant make-up never obscures the pathetic emotions.
Spectrum   VHS, Beta   B

The Hunger

US   1983   93m   ●   colour

Catherine Deneuve, David Bowie, Susan Sarandon, Cliff de Young
D, Tony Scott

Chic vampire yarn. Deneuve is supposed to be 200 years old but looks splendid; her fellow vampire, Bowie, spends most of the time in a box. The director's background in commercials shows.
MGM/UA   VHS, Beta   A

I Walked with a Zombie

US   1943   66m   ●   bw

Frances Dee, James Ellison, Tom Conway, Christine Gordon
D, Jacques Tourneur

As with most of producer Val Lewton's work, the cheap title belies an imaginative film about voodoo antics in the Caribbean. The poetic atmosphere transcends some wooden acting. With: *The Crackup* (qv)
Kingston   VHS, Beta   A

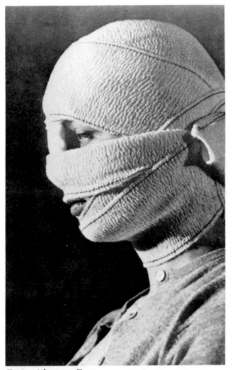

Eyes without a Face

Hunchback of Notre Dame

I Walked with a Zombie

Isle of the Dead*
US 1945 69m ● bw

Boris Karloff, Ellen Drew, Jason Robards, Alan Napier
D, Mark Robson

Erratic but haunting piece from the Lewton stable is set on a Greek island marooned by war and bedevilled by plague. Karloff shines as a Greek general; includes a splendid premature burial. With: *Berlin Express* (qv)
Kingston VHS, Beta A

Lust for a Vampire
GB 1970 92m ● colour

Ralph Bates, Michael Johnson, Barbara Jefford, Suzanna Leigh
D, Jimmy Sangster

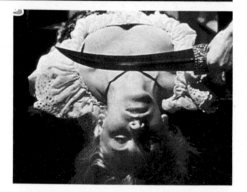

To build a finishing school in the grounds of Karnstein Castle, stronghold of vampires, is asking for trouble; it duly arrives. Workmanlike piece, with good performances and a powerfully sensual atmosphere.
Thorn EMI VHS, Beta A

Magic*
US 1978 106m ● colour

Anthony Hopkins, Ann-Margret, Burgess Meredith, Ed Lauter, E J Andre
D, Richard Attenborough

Slightly laborious variation by an unlikely director on the theme of the ventriloquist who takes his dummy seriously. Hopkins is the psychotic ventriloquist; Ann-Margret the childhood idol caught up in his madness.
Intervision VHS, Beta, V2000 A

The Mephisto Waltz*
US 1971 105m ● colour

Alan Alda, Jacqueline Bisset, Curt Jurgens, Barbara Parkins
D, Paul Wendkos

The rousing tale of a diabolic classical pianist (Jurgens) who bequeathes his soul to a journalist (a pre-*MASH* Alan Alda). Enjoyable hocus-pocus, brought off with the director's characteristic style and swagger.
CBS/Fox VHS, Beta, Laser A

The Monster (It's Alive)*
US 1973 91m ● colour

John Ryan, Sharon Farrell, Andrew Duggan, Guy Stockwell, James Dixon
D, Larry Cohen

A mutant killer is born to Los Angeles parents. John Ryan's deeply-felt performance as the father would have graced a much worthier film; some inspired moments from the director.
Rank VHS, Beta A

The Nightcomers
GB 1971 96m ● colour

Stephanie Beacham, Marlon Brando, Thora Hird, Harry Andrews, Verna Harvey
D, Michael Winner

The formative experiences of the two evil children from Henry James's story, *The Turn of the Screw*. A clever idea, weakened by erratic direction and Brando's eccentric performance as the weird gardener.
Embassy VHS, Beta, V2000, Laser A

Nosferatu**
Germany 1921 63m ● bw

Max Schreck, Gustav Von Wangenheim, Greta Schroeder, Alexander Granach
D, F W Murnau

A pirated version of Bram Stoker's *Dracula* novel, it led to an action for breach of copyright and the destruction of official prints. Creaky, but notable for the director's strong visual sense and Schreck's haunting Count Orlok.
Thorn EMI VHS, Beta C

Nosferatu the Vampire*
WG/France 1979 107m ● col

Klaus Kinski, Isabelle Adjani, Bruno Ganz
D, Werner Herzog

Reverential re-make of Murnau's classic by the perversely talented Herzog. Kinski's Count cannot match Schreck's original for sheer, terrifying ugliness but the film is visually elegant and deliciously spooky.
CBS/Fox VHS, Beta A

The Omen*
US 1976 111m ● colour

Gregory Peck, Lee Remick, David Warner, Billie Whitelaw, Leo McKern
D, Richard Donner

Gregory Peck plays the US Ambassador to London, burdened with an anti-Christ for a son. Smartly mounted diabolical thriller which improves on its inspiration, *The Exorcist* (qv), by not taking itself too seriously.
CBS/Fox VHS, Beta, Laser A

The Phantom of the Opera**
US   1926   74m   ●   bw

Lon Chaney, Mary Philbin, Norman Kerry, Gibson Gowland
D, Rupert Julian

Lon Chaney's skills are well displayed in this stylish silent version of Gaston Leroux's novel about the deformed musician luring a prima donna to his nook beneath the Paris Opera. Chilling moments; fine backstage atmosphere.
Spectrum   VHS, Beta   B

The Pit and the Pendulum*
US   1961   85m   ●   colour

Vincent Price, Barbara Steele, John Kerr
D, Roger Corman

Corman taking on Edgar Allan Poe with a low budget but lots of visual imagination. No great shakes as drama but marvellous on feel and imagery; with raging storms, crumbling castles and spooky torture chambers.
Guild   VHS, Beta, V2000   A

Poltergeist*
US   1982   114m   ●   colour

Jobeth Williams, Craig T Nelson, Beatrice Straight, Dominique Dunne
D, Tobe Hooper

Cosy American suburbia is shattered by poltergeists leaking out of a television set. Producer Steven Spielberg and director Hooper seem to have different styles in mind. They fuse best in the early scenes, full of dark humour.
MGM   VHS, Beta, V2000, Laser   A

The Possession of Joel Delaney
US   1971   104m   ●   colour

Shirley Maclaine, Perry King, Lisa Kohane, David Ellacott
D, Waris Hussein

A serious political message lurks inside this bizarre film, though the surface is dominated by the usual tricks of occult thrillers. Maclaine is a divorcée confronted by decapitation, exorcism etc.
Precision   VHS, Beta   A

The Premature Burial*
US   1961   81m   ●   colour

Ray Milland, Heather Angel, Hazel Court, Richard Ney, Alan Napier
D, Roger Corman

A man's fear of incarceration proves only too justified but he exacts full revenge. Milland proves a jarring replacement for Corman's usual lead, Vincent Price, but the director's ghoulish style prevails.
Guild   VHS, Beta, V2000   A

See it the way it was originally made! Uncut! Every scene intact!

Psycho***
US 1960 108m ● bw

Anthony Perkins, Janet Leigh, John Gavin, Vera Miles, John McIntyre
D, *Alfred Hitchcock*

Psycho is one of the supreme examples of Hitchcock's ability to manipulate an audience. A tribute to his mastery is that the effect is hardly less powerful on subsequent viewings when the spectator knows what is going to happen. All through the film Hitchcock plants expectations in the audience's mind which turn out to be disconcertingly false. At the start Marion (Janet Leigh) makes off with $40,000 entrusted to her by her boss; she has a nasty movement when she is questioned by a cop. But what seems like a simple exercise in suspense takes on an entirely different dimension when Marion stops at the motel run by the sinister Norman Bates (Perkins). Hitchcock has the effrontery to kill off his leading lady only a third of the way through the picture. The shower murder is one of the cinema's most horrific scenes, yet the effect is conveyed almost entirely by suggestion: a knife, a scream, a trickle of blood. It is the audience that puts the images together. The scene was a complicated technical exercise, lasting only 45 seconds on the screen but taking seven days to shoot. Hitchcock has said of *Psycho:* "I don't care about the subject matter; I don't care about the acting; but I do care about the pieces of film and the soundtrack and all the technical ingredients that make the audience scream". When *Psycho* first came out, the critics attacked it for poor taste. The public disagreed and within a few years a film that had cost the modest sum of $800,000 had earned more than $15m.

CIC VHS, Beta A

Psycho II*
US   1983   109m   ●   colour

Anthony Perkins, Vera Miles, Meg Tilly,
Robert Loggia, Dennis Franz
D, Richard Franklin

Brave, belated sequel to Hitchcock's classic
(qv), with Anthony Perkins returning to the
fateful motel after 22 years in the asylum.
Routine slaughter ultimately takes over but
there are passages to relish.
CIC   VHS, Beta   A

Rabid*
Canada   1976   84m   ●   colour

Marilyn Chambers, Joe Silver, Howard
Ryshpan, Patricia Cage, Susan Roman
D, David Cronenberg

After some experimental plastic surgery, a
bike crash victim (Chambers) develops a
blood lust and starts spreading rabies. An
outrageous, witty and compelling piece of
nastiness from an inimitable director.
Intervision   VHS, Beta, V2000   A

Repulsion**
GB   1965   104m   ●   bw

Catherine Deneuve, Ian Hendry, John
Fraser, Patrick Wymark
D, Roman Polanski

Polanski's remarkable British debut, with
Deneuve as a Belgian manucurist barri-
cading herself in a creepy London flat and
descending into madness and murder. A
masterly study of human disintegration;
not for the faint-hearted.
Videomedia   VHS, Beta, V2000   B

The Shining*
GB   1980   143m   ●   colour

Jack Nicholson, Shelley Duvall, Danny
Lloyd, Barry Nelson, Scatman Crothers
D, Stanley Kubrick

Nicholson becomes the caretaker of a
remote Colorado hotel, knowing that it
drove his predecessor beserk. Indulgent,
disappointing excursion into horror territory by a heavyweight director, with suitably chilling moments.
Warner   VHS, Beta   A

A Study in Terror*
GB   1965   95m   ●   colour

John Neville, Donald Houston, John
Fraser, Anthony Quayle
D, James Hill

How Sherlock Holmes tracked down Jack
the Ripper. Rather muted treatment of a
good idea; the compensations are a nicely
atmospheric evocation of Victorian London
and some sterling character work from the
strong cast.
Videomedia   VHS, Beta, V2000   A

Suspiria

Italy 1976 93m ● colour

Jessica Harper, Alida Valli, Joan Bennett, Stefania Casini, Udo Kier
D, Dario Argento

A touch of *The Exorcist* (qv) as a young American girl turns up at a sinister German dance academy once the home of a notorious witch. The absurdities of the plot are masked by the director's stylistic flair and a strident rock score.
Thorn EMI VHS, Beta A

Tales of Terror*

US 1962 90m ● colour

Vincent Price, Peter Lorre, Basil Rathbone, Debra Paget
D, Roger Corman

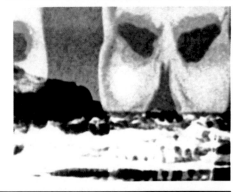

Three stories of Edgar Allan Poe, intelligently adapted by scriptwriter Richard Matheson and brought to life with the customary Corman style. Price and Lorre let their hair down; old-timer Rathbone plays it straight.
Guild VHS, Beta, V2000 A

The Thing

US 1982 107m ● colour

Kurt Russell, A Wilford Brimley, T K Carter, David Clennon, Richard Dysart
D, John Carpenter

Horror crossed with sci-fi in a partial remake of the 1951 film, *The Thing From Another World* (qv). Aliens at large in the Antarctic cause mayhem and murder, orchestrated by Carpenter with less than his usual flair.
CIC VHS, Beta, V2000, Laser A

To the Devil a Daughter

GB/WG 1976 90m ● colour

Richard Widmark, Christopher Lee, Denholm Elliott, Honor Blackman, Nastassia Kinski
D, Peter Sykes

Occult novelist Widmark tries to protect vulnerable Kinski from Lee and his satanists. Adaptation of Dennis Wheatley and a worthy last film for the cinema by the famous Hammer company.
Thorn EMI VHS, Beta A

Whatever Happened to Aunt Alice?*

US 1969 98m ● colour

Geraldine Page, Ruth Gordon, Rosemary Forsyth, Robert Fuller
D, Lee H Katzin

Page as an impoverished widow trying to keep her head above water by bumping off her rich housekeepers. Entertaining exercise in bloodletting, played to the hilt and adroitly handled by the director.
Rank VHS, Beta A

Whatever Happened to Baby Jane?*

US 1962 129m ● bw

*Bette Davis, Joan Crawford, Victor
Buono, Anna Lee, Marjorie Bennett
D, Robert Aldrich*

Davis plays a demented former child-star
wreaking vengeance on crippled sister
Crawford in a tumbledown Hollywood
mansion. Glorious piece of nonsense, with
virtuoso performances.
Warner VHS, Beta A

The Wicker Man*

GB 1973 83m ● colour

*Edward Woodward, Britt Ekland,
Christopher Lee, Ingrid Pitt
D, Robin Hardy*

Anthony (*Sleuth*) Shaffer scripted this
unassuming tale of devilish happenings on
a remote Scottish island. Released as the
lower half of a double bill, it is now a cult
piece. Woodward is the policeman sent to
investigate.
Thorn EMI VHS, Beta A

Witchfinder General*

GB 1968 87m ● colour

*Vincent Price, Rupert Davies, Ian Ogilvy,
Patrick Wymark, Hilary Dwyer
D, Michael Reeves*

Price in one of his more restrained moods
as a vicious lawyer roaming Cromwell's
England on the lookout for witches. Intelligently written, impressively photographed
film by a promising young director who died
soon afterwards.
Hokushin VHS, Beta A

Wolfen

US 1981 114m ● colour

*Albert Finney, Diane Venora, Edward
James Olmos, Gregory Hines
D, Michael Wadleigh*

Something very nasty is at large on the
streets of New York and detective Albert
Finney is put on the case. Heavy-going
piece, with allegorical pretensions, which
would have done better to stick to the gore.
Warner VHS, Beta A

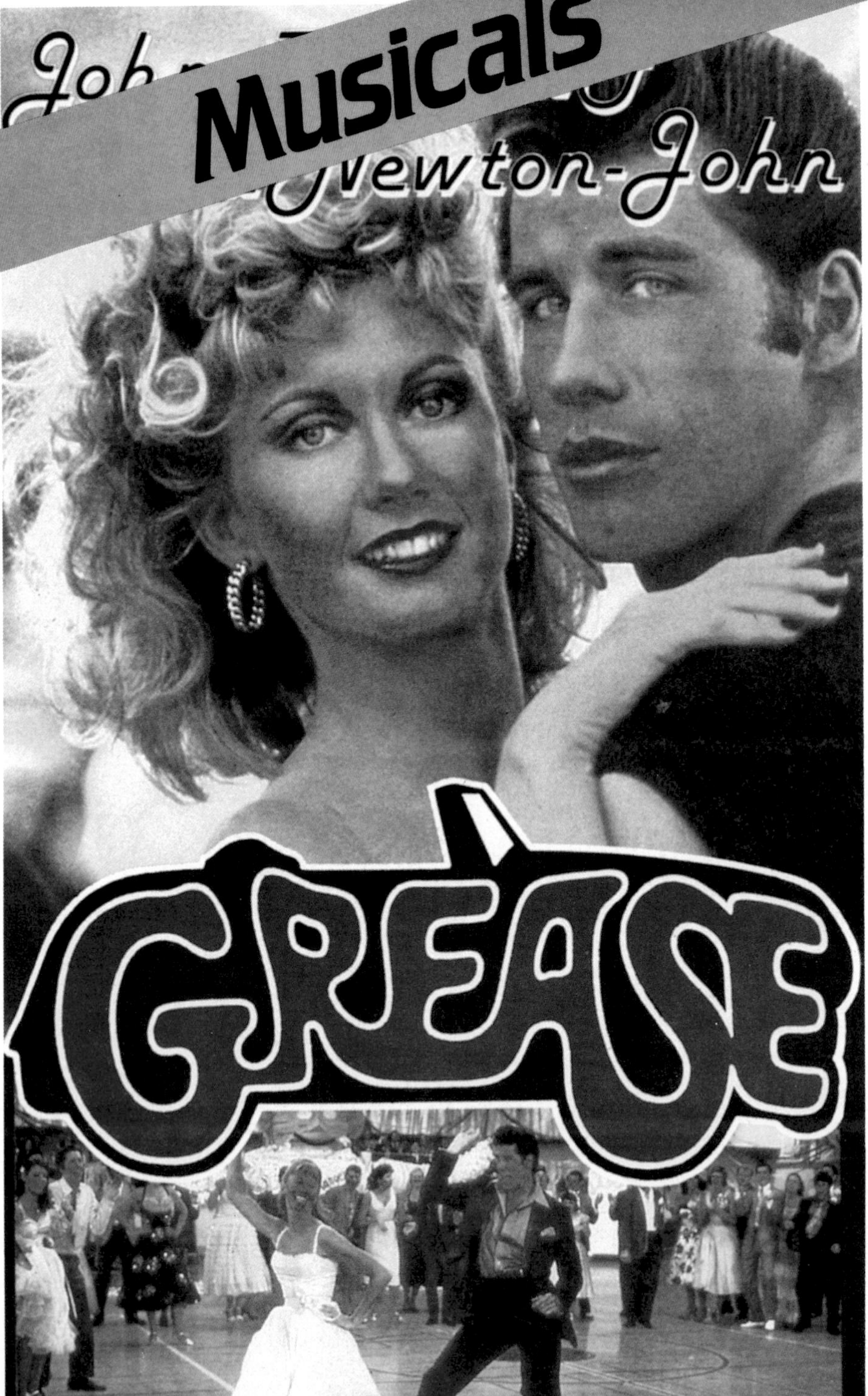
Newton-John
GREASE

All That Jazz*
US 1979 123m ● colour

Roy Scheider, Jessica Lange, Ann Reinking, Leland Palmer, Ben Vereen
D, Bob Fosse

Scheider gives a thinly disguised portrait of director Fosse, a choreographer/film-maker, driving himself towards a heart attack with women, drugs and overwork. Some stylish numbers, but musicals do not come much blacker than this.
CBS/Fox   VHS, Beta, V2000, Laser   A

Anchors Aweigh*
US 1945 139m □ colour

Frank Sinatra, Gene Kelly, Kathryn Grayson, Jose Iturbi, Sharon McManus
D, George Sidney

Sailors Sinatra and Kelly take shore leave in Hollywood; Sinatra sings "I Fall in Love Too Easily" and Kelly hoofs it with Jerry, the cartoon mouse – one of the first successful mixes of animation and live action.
MGM/UA   VHS, Beta   B

Annie*
US 1982 128m □ colour

Albert Finney, Carol Burnett, Aileen Quinn, Ann Reinking, Bernadette Peters
D, John Huston

Nine-year-old Aileen Quinn as little orphan Annie, melting the heart of billionaire Finney. Appealing performances carry through this first foray into the musical by veteran director Huston.
RCA/Columbia   VHS, Beta, CED   A

The Band Wagon**
US 1953 108m □ colour

Fred Astaire, Jack Buchanan, Oscar Levant, Cyd Charisse, Nanette Fabray
D, Vincente Minnelli

A band wagon definitely to be jumped on, as song-and-dance-man Astaire, temperamental producer Buchanan and leggy ballerina Charisse get together to stage a Broadway show. Songs include "Louisiana Hayride", "Dancing in the Dark".
MGM/UA   VHS, Beta, Laser   B

The Barkleys of Broadway*
US 1949 105m □ colour

*Fred Astaire, Ginger Rogers, Oscar
Levant, Jacques François, Billie Burke
D, Charles Walters*

Fred and Ginger reunited after nine years,
as a musical comedy duo who split up
when she wants to become a straight
actress. A sequel to *Easter Parade* (qv);
Judy Garland, the original lead, withdrew
through illness.
MGM/UA VHS, Beta B

The Beggar's Opera*
GB 1952 90m □ colour

*Laurence Olivier, Stanley Holloway,
Dorothy Tutin, Daphne Anderson
D. Peter Brook*

First excursion into cinema by a very young
Peter Brook has Laurence Olivier singing
and dancing his way through eighteenth
century London as highwayman Macheath
in a spirited version of John Gay's 1728
opera. An engaging oddity.
Thorn EMI VHS, Beta B

The Best Little Whorehouse in Texas
US 1982 110m v colour

*Burt Reynolds, Dolly Parton, Charles
Durning, Dom DeLuise, Jim Nabors
D, Colin Higgins*

Sheriff Burt Reynolds joins brothel-keeper
Dolly Parton in the fight against a clean-up
campaign by a TV consumer champion.
They are also fighting a coy script and
lacklustre songs.
CIC VHS, Beta A

The Boy Friend
GB 1971 125m □ colour

*Twiggy, Christopher Gable, Max Adrian,
Tommy Tune, Glenda Jackson
D, Ken Russell*

Sandy Wilson's affectionate 1920s
pastiche turned upside down by Russell
trying to be Busby Berkeley; despite the
stylistic mess and some over-acting, the
jollity and Twiggy's naïve charm work
surprisingly well.
MGM/UA VHS, Beta A

Brigadoon
US 1954 108m ☐ colour

*Gene Kelly, Cyd Charisse, Van Johnson,
Jimmy Thompson, Elaine Stewart
D, Vincente Minnelli*

The Alan Jay Lerner/Frederick Loewe
whimsy about two Americans discovering
a magical village in Scotland which comes
to life only once a century. The film is
agreeable, although some of the magic is
missing.
MGM/UA VHS, Beta B

Bugsy Malone**
GB 1976 93m ☐ colour

*Scott Baio, Jodie Foster, Florrie Digger,
John Cassisi, Martin Lev
D, Alan Parker*

Ingenious musical spoof on the American
gangster movie of the 1930s with all the
parts played by kids. Writer/director Parker
gets full marks for originality and his young
actors relish every minute. A treat for all
ages.
Rank VHS, Beta, Laser A

Cabaret**
US 1972 120m ● colour

*Liza Minnelli, Joel Grey, Michael York,
Helmut Griem, Marisa Berenson
D, Bob Fosse*

Style and verve from Minnelli as Isher-
wood's Sally Bowles, a Berlin nightclub
singer of the 1930s; Grey's lewdly arch MC
gained him an Oscar. Director Fosse triumphantly translates a stage musical into
brilliant, biting cinema.
Rank VHS, Beta, Laser A

Camelot
US 1967 166m ☐ colour

*Richard Harris, Vanessa Redgrave, David
Hemmings, Lionel Jeffries, Franco Nero
D, Joshua Logan*

The Lerner/Loewe version of King Arthur
and Guenevere with splendid sets and
costumes (both of which won Oscars). But
Logan seems to think he is still directing for
the theatre and neither of the principals are
strong singers.
Warner VHS, Beta A

Carefree*
US 1938 80m ☐ bw

*Fred Astaire, Ginger Rogers, Ralph
Bellamy, Luella Gear, Clarence Kolb
D, Mark Sandrich*

A minor piece by the highest Astaire-
Rogers standards, but the stars are always
watchable and there is a good sprinkling of
Irving Berlin numbers. Fred is a psychia-
trist; Ginger one of his clients. With: *Easy
Living*, 1949 drama.
Kingston VHS, Beta A

A Damsel in Distress*
US 1937 87m ☐ bw

Fred Astaire, George Burns, Gracie Allen, Joan Fontaine
D, George Stevens

Young Joan Fontaine pluckily fills in for Ginger Rogers as a rich heiress living in a castle, mistaken by Astaire for a chorus girl. Story by P G Wodehouse, songs by George and Ira Gershwin. With: *Old Man Rhythm*, 1935 musical.
Kingston VHS, Beta A

Doctor Dolittle
US 1967 152m ☐ colour

Rex Harrison, Anthony Newley, Samantha Eggar, Richard Attenborough, Peter Bull
D, Richard Fleischer

Rex Harrison does his polished best in this high-budget version of Hugh Lofting's stories of the doctor who talks to animals in their own language and sets off in search of the Great Pink Sea Snail. Rather long.
CBS/Fox VHS, Beta, Laser A

Easter Parade**
US 1948 103m ☐ colour

Fred Astaire, Judy Garland, Ann Miller, Peter Lawford, Clinton Sundberg
D, Charles Walters

Not much of a plot but one is hardly needed with such exuberant performers and 17 Irving Berlin numbers, including "Drum Crazy", "Shakin' the Blues Away", "We're a Couple of Swells" and the title song.
MGM/UA VHS, Beta, Laser B

Expresso Bongo
GB 1959 109m v bw

Laurence Harvey, Sylvia Sims, Yolande Donlan, Cliff Richard, Meier Tzelniker
D, Val Guest

Mildly satirical look at the 1950s British pop scene of coffee bars and rock 'n' roll, with Harvey as a pushy agent trying to make teenager Cliff Richard into a star. Interesting period piece.
Videomedia VHS, Beta, V2000 A

Fame*
US 1980 133m v colour

Irene Cara, Lee Curreri, Laura Dean, Paul McCrane, Gene Anthony Ray
D, Alan Parker

Young hopefuls aim for stardom at Manhattan's High School for the Performing Arts. This energetic mixture of realism and fantasy became a TV series; but the frequent use of four-letter words makes it unsuitable for younger children.
MGM/UA   All systems   A

Fiddler on the Roof*
US   1971   181m   □   colour

Topol, Norma Crane, Leonard Frey, Molly Picon, Paul Mann
D, Norman Jewison

Stolid, respectful version of the Jerry Bock/Sheldon Harnick stage hit about a simple Russian-Jewish milkman trying to find husbands for his five daughters. Topol's lively and sympathetic portrait transcends some dull bits.
Warner   VHS, Beta, V2000   A

Flashdance
US   1983   91m   v   colour

Jennifer Beals, Michael Nouri, Lilia Skala, Sunny Johnson, Kyle T Heffner
D, Adrian Lyne

Cinderella story of a young female welder who does sexy dances in a nightclub but really wants to become a ballerina. Strident tribute to the aerobics cult, in which Albinoni and Debussy share the tracks with electronic funk.
CIC   VHS, Beta   C

42nd Street***
US   1933   89m   □   bw

Warner Baxter, Ruby Keeler, Bebe Daniels, Dick Powell, Ginger Rogers
D, Lloyd Bacon

Never mind that the putting-on-a-show story had been used before and has been repeated many times since; this is the film that did it best, helped by the dazzlingly inventive routines of the master choreographer, Busby Berkeley.
Warner   VHS, Beta, CED   A

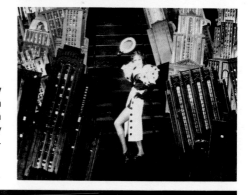

Gigi**
France   1958   116m   □   colour

Leslie Caron, Louis Jourdan, Maurice Chevalier, Hermione Gingold
D, Vincente Minnelli

Stylish, elegant and enchanting musical version of the Colette story about a Parisian schoolgirl who marries a rake; Cecil Beaton sets plus indestructible Lerner/Loewe numbers, including Chevalier's "Thank Heaven for Little Girls".
MGM/UA   VHS, Beta, Laser   B

Carefree

Fame

Hallelujah, I'm a Tramp

Les Girls

US 1957 113m ☐ colour

Gene Kelly, Kay Kendall, Mitzi Gaynor,
Taina Elg, Leslie Phillips
D, George Cukor

Kay Kendall as one of a trio of showgirls looking back over their past and outshining the others with her polished playing. "Woman's director" Cukor supplies his usual gloss, but the Cole Porter numbers are not particularly memorable.
MGM/UA VHS, Beta B

Grease*

US 1978 105m ☐ colour

John Travolta, Olivia Newton-John,
Stockard Channing, Eve Arden
D, Randal Kleiser

The most successful musical of the 1970s, it shrewdly combines 1950s high-school nostalgia with the star-appeal of the strutting disco kid, John Travolta. A sprinkling of Hollywood veterans is added for the mums and dads.
CIC VHS, Beta, V2000, Laser C

The Great Caruso*

US 1951 109m v bw

Mario Lanza, Ann Blyth, Dorothy Kirsten,
Jarmila Novotna, Carl Benton Reid
D, Richard Thorpe

Glamorised rags-to-riches biopic of Enrico Caruso, the poor boy from Naples who became the darling of the world's opera houses. No great shakes as drama, but the singing is the thing and Lanza has both the voice and the charisma.
MGM/UA VHS, Beta B

Guys and Dolls*

US 1955 149m v colour

Frank Sinatra, Marlon Brando, Jean
Simmons, Vivian Blaine, Stubby Kaye
D, Joseph L Mankiewicz

Frank Loesser's musical venture into Damon Runyon country is given the heavy studio treatment by the director. Brando is not everyone's idea of Sky Masterson, but the dolls (Simmons, Blaine) are fine and the Runyon flavour seeps through.
CBS/Fox VHS, Beta A

Hallelujah, I'm a Tramp*

US 1933 77m v bw

Al Jolson, Harry Langdon, Madge Evans,
Frank Morgan, Chester Conklin
D, Lewis Milestone

Depression-era piece by Rodgers and Hart (both appear in small parts). Jolson, a Central Park squatter, falls in love with a girl who has lost her memory. The novelty is that the dialogue is mainly in recitative.
Intervision VHS, Beta A

High Society*
US 1956 107m v colour

Bing Crosby, Frank Sinatra, Grace Kelly,
Celeste Holm, Louis Armstrong
D, Charles Walters

Bing turns up at ex-wife Kelly's wedding and woos her back. Musical re-make of *The Philadelphia Story* with some of the wit lost; the Cole Porter songs ("Who Wants to be a Millionaire?", "Well, Did You Evah?") help to keep it moving.
MGM/UA VHS, Beta C

The Jazz Singer*
US 1980 110m v colour

Neil Diamond, Laurence Olivier, Lucie
Arnaz, Catlin Adams, Sully Boyar
D, Richard Fleischer

The Al Jolson hit of 1927, which has gone down in cinema history as the first talkie, remade as an enjoyable vehicle for Neil Diamond. A cantor's boy is torn between family and showbiz; good songs, lashings of sentiment.
Thorn EMI VHS, Beta A

Jesus Christ Superstar
US 1973 103m v colour

Ted Neeley, Carl Anderson, Yvonne
Elliman, Barry Dennen
D, Norman Jewison

The Tim Rice/Andrew Lloyd Webber "rock opera" about young visitors to the Holy Land acting out the last days of Christ. A self-conscious screen treatment by director Jewison and it works only in parts.
CIC VHS, Beta, Laser C

The King and I*
US 1956 128m □ colour

Deborah Kerr, Yul Brynner, Rita Moreno,
Martin Benson, Alan Mowbray
D, Walter Lang

Deborah Kerr (dubbed by Marni Nixon) and Brynner, using his own voice, work through a marvellous Rodgers and Hammerstein score as an English governess and the Siamese king; director Lang is content to let them get on with it.
CBS/Fox VHS, Beta, Laser A

Kiss Me Kate*
US 1953 105m v colour

Howard Keel, Kathryn Grayson, Ann
Miller, Keenan Wynn, Bobby Van
D, George Sidney

By Cole Porter out of Shakespeare's *The Taming of the Shrew* with Keel and Grayson as the warring partners, offstage as well as on. Lively, faithful version of the stage hit and one of the few musicals to be shot in 3D.
MGM/UA VHS, Beta B

A Little Night Music
Austria/WG 1977 102m v col

Elizabeth Taylor, Diana Rigg, Len Cariou,
Lesley-Anne Down, Hermione Gingold
D, Harold Prince

Fine Stephen Sondheim musical, from
Ingmar Bergman's *Smiles of a Summer*
Night, makes a bumpy transition to the
screen. It switches locale from Sweden to
Vienna, loses six songs and miscasts Liz
Taylor.
Guild   VHS, Beta, V2000   A

Meet Me in St Louis***
US 1944 108m □ colour

Judy Garland, Margaret O'Brien, Tom
Drake, Leon Ames, Mary Astor
D, Vincente Minnelli

Lyrical and affectionate portrait of family
life in small-town America at the turn of the
century. Decoratively staged and with
some enduring numbers, including "Have
Yourself a Merry Little Christmas" and
"The Trolley Song".
MGM/UA   VHS, Beta, Laser   B

My Fair Lady**
US 1964 164m □ colour

Rex Harrison, Audrey Hepburn, Stanley
Holloway, Gladys Cooper, Jeremy Brett
D, George Cukor

Elegant but literal transcription of the
Lerner/Loewe version of Shaw's *Pyg-*
malion, with director Cukor happy to let the
performances take over. Harrison and
Holloway repeat their stage roles; Hepburn
was dubbed by Marni Nixon.
CBS/Fox   VHS, Beta, Laser   A

New York, New York**
US 1977 153m v colour

Liza Minnelli, Robert de Niro, Lionel
Stander, Barry Primus
D, Martin Scorsese

Bitter-sweet romance between saxophonist
De Niro and aspiring singer Minnelli set in
the Big Band era of the 1940s. Both a
tribute to the traditional studio-bound
Hollywood musical and a shrewd re-
working of its conventions.
Warner   VHS, Beta   A

Oklahoma!*
US 1955 140m □ colour

Gordon Macrae, Shirley Jones, Rod
Steiger, Gloria Grahame
D, Fred Zinnemann

Stiffly directed version of the great Rodgers
and Hammerstein hymn to rural America.
It hardly matters with songs like "Oh, What
a Beautiful Mornin'", "The Surrey With
the Fringe on Top" and "People Will Say
We're in Love".
CBS/Fox   VHS, Beta, Laser   A

Oliver!*
GB 1968 140m □ colour

Ron Moody, Oliver Reed, Harry Secombe,
Mark Lester, Jack Wild
D, Carol Reed

This first musical by veteran director Reed
won him an Oscar. An efficient rendering
of Lionel Bart's jolly version of *Oliver Twist*;
impressive sets from John Box and young
Mark Lester makes a fetching Oliver.
RCA/Columbia VHS, Beta R

On the Town***
US 1949 94m □ colour

Gene Kelly, Frank Sinatra, Jules
Munshin, Vera-Ellen, Betty Garrett, Ann
Miller
D, Gene Kelly and Stanley Donen

Simple tale of three sailors on 24-hour
shore leave in New York and the girls they
chum up with. But what a musical!
Exhilarating, inventive, six smashing performances; apt and great numbers.
MGM VHS, Beta, Laser B

Paint Your Wagon
US 1969 157m v colour

Lee Marvin, Clint Eastwood, Jean
Seberg, Harve Presnell, Ray Walston
D, Joshua Logan

Lumbering and bizarrely cast treatment for
the Lerner-Loewe piece about gold prospectors in California. The compensation is
Lee Marvin, doing his drunk act and
treating us to a suitably abrasive interpretation of "Wand'rin Star".
Arena VHS, Beta B

Pennies from Heaven
US 1982 103m ● colour

Steve Martin, Bernadette Peters,
Christopher Walken, Jessica Harper,
John McMartin
D, Herbert Ross

Dennis Potter transplanted his dreams-
versus-reality tale of a 1930s sheet-music
salesman from Britain to America and
gained in gloss what he lost in bite. Some
will prefer the original British TV version.
MGM/UA VHS, Beta A

Pink Floyd The Wall
GB   1982   91m   ●   colour

Bob Geldof, Christine Hargreaves, James Laurenson, Eleanor David, Bob Hoskins
D, Alan Parker

Geldof as a burnt-out rock star no longer able to distinguish fantasy from reality; an acclaimed album by Pink Floyd; animated sequences from cartoonist Gerald Scarfe. Director Parker frenetically tries to make it all cohere.
Thorn EMI   VHS, Beta   A

Saturday Night Fever*
US   1977   114m   v   colour

John Travolta, Karen Lynn Gorney, Barry Miller, Joseph Cali, Paul Pape
D, John Badham

Sultry John Travolta rocketed to stardom as the king of the Brooklyn discos. A slick and pacy concoction, making up in energy what it lacks in style, and accurately evoking youthful revolt against suburban boredom.
CIC   VHS, Beta, Laser   C

Seven Brides for Seven Brothers**
US   1954   100m   □   colour

Howard Keel, Jane Powell, Jeff Richards, Russ Tamblyn, Tommy Rall
D, Stanley Donen

Fast-moving and high-spirited romp in 1850 Oregon as the country lads make off with girls from the local town; with a celebrated barn-building sequence created by choreographer Michael Kidd.
MGM/UA   VHS, Beta, Laser, CED   B

Show Boat*
US   1951   104m   v   colour

Kathryn Grayson, Howard Keel, Ava Gardner, William Warfield, Joe E Brown
D, George Sidney

All aboard for a vigorous paddle down the Mississippi, with Keel as the gambler Ravenal, Grayson the shy Magnolia; the rich Jerome Kern/Oscar Hammerstein score includes "Can't Help Lovin' Dat Man" and "Old Man River".
MGM/UA   VHS, Beta   B

Silk Stockings*
US   1957   111m   □   colour
Fred Astaire, Cyd Charisse, Peter Lorre,
Janis Paige, George Tobias
D, Rouben Mamoulian

Stylish musical re-working of the old Garbo
vehicle, *Ninotchka*, with exquisite dancing
and a delicate use of colour. Charisse plays
a frosty Russian commissar and Astaire is
just the man to melt her.
MGM/UA   VHS, Beta, Laser   B

Singin' in the Rain***
US   1952   102m   cert   □   colour
Gene Kelly, Donald O'Connor, Debbie
Reynolds, Jean Hagen
D, Gene Kelly and Stanley Donen

Like many of the great MGM musicals of
the 1940s and early 1950s, *Singin' in the
Rain* was not taken from an established
stage show but created specially for the
cinema. It took the cinema as its subject;
the arrival of the talkies in Hollywood in
1927 and the threat this posed to the
career of a leading silent star whose
squeaky voice would sink her as soon as
the audience heard it. Jean Hagen plays
her to the hilt; Kelly is her screen partner;
and Debbie Reynolds is the young actress
who saves the star's first sound film from
disaster by dubbing her voice. As a story,
sharp and witty, it would stand on its own
without music: the sparkling numbers
(lyrics by Arthur Freed, the film's producer;
music by Nacio Herb Brown) come as a
magnificent bonus. The best known is the
title song, interpreted by a joyous Kelly
sploshing his way down an empty street
"just singin' and dancin' in the rain". But
almost as memorable is Donald O'Connor's "Make 'em Laugh", Kelly and Reynolds on an empty stage with "You Were
Meant For Me" and the infectiously opti-
mistic "Good Morning", performed by
Kelly, O'Connor and Reynolds. There is
also the elaborately staged "Broadway
Ballet", which has little to do with the rest
of the film but is an excuse to bring in Cyd
Charisse: and why not? Made by virtually
the same team that was responsible for the
exhilarating *On the Town* (qv), *Singin' in
the Rain* is a film of constant delight that
never loses its freshness and bounce.
Sadly, it is the sort of picture that could not
be made today, when no musical is risked
on film that has not already proved itself in
the theatre.
MGM/UA   VHS, Beta   B

The Sky's the Limit*
US 1943 87m ☐ bw

*Fred Astaire, Joan Leslie, Robert
Benchley, Robert Ryan
D, Edward H Griffith*

Astaire plays an airman who falls in love
with a news photographer (Leslie). Slight
but enjoyable; the Harold Arlen/Johnny
Mercer numbers include "My Shining
Hour" and "One For My Baby". With: *Step
Lively*, 1944 musical.
Kingston VHS, Beta A

The Sound of Music***
US 1965 167m ☐ colour

*Julie Andrews, Christopher Plummer,
Eleanor Parker, Richard Haydn
D, Robert Wise*

Though the critics tended to be sniffy about
it, *The Sound of Music* quickly became the
most successful film musical ever; for a
time it was the most successful film of any
type, beating the record long held by *Gone
With the Wind*. Some delighted customers
saw it dozens of times; it was a phenomenon and yet it was an unexpected
success, although it had run for four years
in New York as a stage show. The taste in
the cynical 1960s seemed for more astringent subjects to set to music than this
bland tale of a postulant who leaves the
abbey to become governess to seven
children of the widowed Baron von Trapp,
marries him, leads the family to safety from
the Nazis and lives happily ever after. Nor,
despite her success in her first film, *Mary
Poppins,* was Julie Andrews anything like a
bankable Hollywood name (she had lost the
leading role in *My Fair Lady* to Aubrey
Hepburn for that reason). These apparent
drawbacks turned out to be among the
film's strongest assets. Audiences did still
want sentiment, romance, escapism; they
did respond to Julie Andrews and her
wholesome charm. They also enjoyed the
lush Austrian scenery; and, whether they
realised it or not, they found themselves
ensnared by a very professional piece of
film making, in which narrative, settings,
and musical numbers were perfectly inte-
grated. It helped that before a single
cinema seat had been sold the Rodgers and
Hammerstein score was known the world
over, almost every song a hit in its own
right; but *The Sound of Music* is a lot more
than the sum of its tunes.
CBS/Fox VHS, Beta, Laser A

Summer Holiday*
GB 1962 88m □ colour

Cliff Richard, Lauri Peters, Melvyn Hayes,
Una Stubbs, Ron Moody
D, Peter Yates

Four likeable and wholesome young mechanics take a double-decker bus on a Continental holiday, picking up girls en route. The songs and the dance numbers are unremarkable but Cliff's boyish charm carries the day.

Thorn EMI   VHS, Beta   A

That's Entertainment*
US 1974 121m □ colour

Fred Astaire, Gene Kelly, Bing Crosby,
James Stewart, Frank Sinatra
D, Jack Haley Jnr

An anthology of highlights from MGM musicals from the 1930s to the 1950s, introduced by many of the stars who appeared in them; an unashamedly nostalgic wallow, with a good mix of rare and familiar items.

MGM/UA   VHS, Beta, CED   A

That's Entertainment Part 2*
US 1976 121m □ colour

Fred Astaire, Gene Kelly
D, Gene Kelly

The enormous success of the above prompted this sequel, introduced by Astaire and Kelly. Generally less interesting clips include scenes from non-musicals featuring Garbo, Jean Harlow and the Marx Brothers among others.

MGM/UA   VHS, Beta   A

Tommy
GB 1975 103m ● colour

Roger Daltrey, Ann-Margret, Oliver Reed,
Elton John, Eric Clapton
D, Ken Russell

Suitably strident interpretation of the rock opera by Pete Townshend and The Who, with Daltrey as the deaf, blind and dumb hero who becomes pinball champion of the world. Gallant performance by Ann-Margret as his mum.

Thorn EMI   VHS, Beta   A

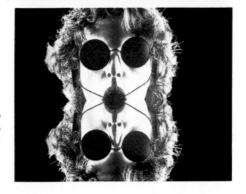

West Side Story**
US 1961 151m v colour

Natalie Wood, Richard Beymer, Russ Tamblyn, Rita Moreno, George Chakiris
D, Robert Wise and Jerome Robbins

Romeo and Juliet transferred to the New York slums, with vibrant choreography from co-director Robbins; energetic performances and excellent Leonard Bernstein/Stephen Sondheim score, including "Tonight" and "America".
Warner VHS, Beta A

Wonderful Life
GB 1964 113m □ colour

Cliff Richard, Walter Slezak, Susan Hampshire, Melvyn Hayes, The Shadows
D, Sidney J Furie

By the time this was made, the Cliff Richard movie formula was wearing thin; but director Furie still managed to put some zip into a story of young pop performers shipwrecked off the Canaries and getting involved with a film crew.
Thorn EMI VHS, Beta A

Yankee Doodle Dandy***
US 1942 126m □ bw

James Cagney, Joan Leslie, Walter Huston, Rosemary de Camp
D, Michael Curtiz

Dynamic impersonation by James Cagney of the composer/song-and-dance man George M Cohan. Shameless piece of patriotism for wartime audiences; songs include "Yankee Doodle Boy", Give My Regards to Broadway", "Over There".
MGM CED only C

The Young Ones
GB 1961 104m □ colour

Cliff Richard, Robert Morley, Carole Gray, Richard O'Sullivan, Melvyn Hayes
D, Sidney J Furie

Cliff and chums put on a show to save a youth club from demolition by property developer Morley. Simple message, delivered with zest; the 19 songs include the title number and the chart-topping "Living Doll".
Thorn EMI VHS, Beta A

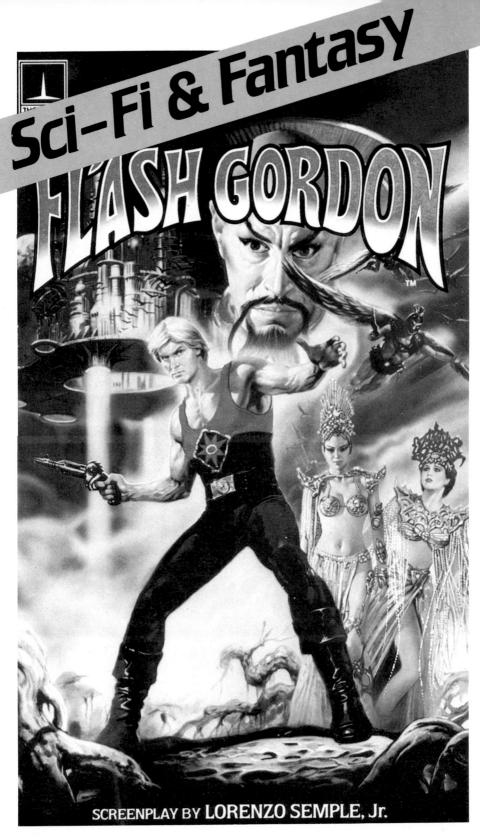

FLASH GORDON

SCREENPLAY BY LORENZO SEMPLE, Jr.

Alien*
GB   1979   111m   ● colour
*Tom Skerritt, John Hurt, Sigourney
Weaver, Veronica Cartwright, Ian Holm
D, Ridley Scott*

Astronaut John Hurt stops off on a planet
and picks up something nasty, which
proceeds to devour his colleagues. Com-
mercially adroit amalgam of *Star Wars* and
The Exorcist (both qv) but the real star is the
monster.
CBS/Fox   VHS, Beta, V2000, Laser   A

Alphaville**
France/Italy   1965   98m   ● col
*Eddie Constantine, Anna Karina, Akim
Tamiroff, Howard Vernon
D, Jean-Luc Godard*

Thinking man's sci-fi in which pulp fiction
hero, Lemmy Caution (Constantine), is set
loose in a bleak, futuristic city ruled by a
giant computer. Godard's apposite alternative title was *Tarzan versus IBM.*
Palace   VHS, Beta   A

Altered States*
US   1980   102m   ● colour
*William Hurt, Blair Brown, Bob Balaban,
Charles Haid
D, Ken Russell*

Harvard boffin Hurt finds he can reverse
the march of evolution and turns himself
into an ape, with explosive results. A novel
variation on a standard horror film formula,
visually inventive with good effects.
Warner   VHS, Beta   A

The Andromeda Strain*
US   1970   124m   ● colour
*Arthur Hill, David Wayne, James Olson,
Kate Reid, Paul Kelly
D, Robert Wise*

Can the scientists save a New Mexico
village from a deadly organism? Director
Wise goes for heavy detail when simple
suspense would have been more effective
but pulls off a gripping climax.
CIC   VHS, Beta   A

Barbarella*
France/Italy   1967   97m   ● col
*Jane Fonda, John Phillip Law, Anita
Pallenberg, David Hemmings, Milo
O'Shea
D, Roger Vadim*

A male chauvinist's delight based on a
popular comic strip. Scantily clad Jane
Fonda gets up to all sorts of naughty things
in a galaxy of the fortieth century. It has a
certain camp charm.
CIC   VHS, Beta   A

Battlestar Galactica
US 1979 125m ☐ colour

*Richard Hatch, Dirk Benedict, Lorne
Greene, Maren Jensen, Herb Jefferson
D, Richard A Colla*

Cut-price *Star Wars* (qv) with commander
Lorne Greene and his crew menaced in
space by the dreaded Cylons. Supporting
cast boasts old-timers Lew Ayres, Ray
Milland and Wilfrid Hyde-White; but the
heroes are the special effects men.
CIC VHS, Beta, V2000 A

The Black Hole
US 1979 94m ☐ colour

*Maximilian Schell, Robert Forster,
Anthony Perkins, Joseph Bottoms
D, Gary Nelson*

Mad scientist Schell is determined to learn
the secrets of a mysterious black hole in
space. A rehash from other sci-fi films and,
as so often, the gadgetry is more interesting than the humans.
Disney VHS, Beta R

Blade Runner*
US 1982 114m v colour

*Harrison Ford, Rutger Hauer, Sean
Young, Edward James Olmos
D, Ridley Scott*

Harrison Ford as a twenty-first century Los
Angeles private eye with a mission to
destroy a group of replicants – robots which
are exact copies of human beings. Decorative, violent and energetic.
Warner VHS, Beta A

The Body Stealers
GB/US 1969 91m c colour

*George Sanders, Maurice Evans, Patrick
Allen, Neil Connery, Hilary Dwyer
D, Gerry Levy*

The stealers come from outer space and
are particularly fond of snatching Nato
parachutists; the key seems to be a strange
girl called Lorna who keeps disappearing. A
worthy cast does its best.
Guild VHS, Beta A

Buck Rogers in the 25th Century*

US 1979 89m v colour

Gil Gerard, Pamela Hensley, Erin Gray, Henry Silva, Tim O'Connor
D, Daniel Haller

Frozen in space for 700 years, our hero returns to Earth after a nuclear holocaust. Amiable and unpretentious updating of a pre-war movie serial, with a genial leading man.
CIC VHS, Beta, Laser A

Close Encounters of the Third Kind*

US 1977 129m v colour

Richard Dreyfuss, François Truffaut, Teri Garr, Melinda Dillon, Gary Guffey
D, Steven Spielberg

Slight story of friendly UFOs descending upon an awed Indiana family woven into box-office magic by director Spielberg and his special effects team. This is the "special edition" with new sequences.
RCA/Columbia VHS, Beta, CED R

The Day of the Triffids

GB 1962 94m ● colour

Howard Keel, Nicole Maurey, Kieron Moore, Janette Scott, Alexander Knox
D, Steve Sekely

John Wyndham's classic spine-chiller about meteorites sowing a deadly crop of giant, man-eating plants. Although undersold by indifferent acting and direction, there are still enough scary moments to keep you watching.
Videospace VHS, Beta A

The Day the Earth Caught Fire*

GB 1961 99m v bw

Edward Judd, Janet Munro, Leo McKern, Arthur Christiansen, Michael Goodliffe
D, Val Guest

Unusually cogent excursion into sci-fi for a British studio. The Earth is knocked out of orbit by nuclear test bangs and gets too close to the sun. Tense and credible.
Intervision VHS, Beta A

Death Race 2000*

US 1975 80m ● colour

David Carradine, Simone Griffeth, Sylvester Stallone, Mary Woronov
D, Paul Bartel

It is the year of the transcontinental road race in which drivers must kill as many pedestrians as possible. Director Bartel supplies a shrewd intelligence and delicious sense of the macabre.
Brent Walker VHS, Beta, V2000 A

Demon Seed
US 1977 90m ● colour

Julie Christie, Fritz Weaver, Gerrit
Graham, Berry Kroeger, Lisa Lu
D, Donald Cammell

Giant computer locks up poor Julie Christie,
rapes her and incubates a child. Presuma-
bly intended as an awful warning about the
perils of the technological age, but must be
taken with a large pinch of salt.
MGM/UA VHS, Beta A

Escape from New York*
US 1981 99m ● colour

Kurt Russell, Lee Van Cleef, Ernest
Borgnine, Donald Pleasence, Isaac Hayes
D, John Carpenter

Desperado Kurt Russell tries to win a
reprieve by spiriting the President of the
United States off Manhattan Island, which
has become a top security prison. Director
Carpenter begins crisply, but gets too
clever.
Embassy VHS, Beta, V2000, Laser A

Flash Gordon*
GB 1980 109m ☐ colour

Sam J Jones, Melody Anderson, Topol,
Max Von Sydow, Brian Blessed
D, Michael Hodges

Lavishly mounted reincarnation of the
comic strip hero of the 1930s, once more
taking on the dastardly Emperor Ming (Von
Sydow); director Hodges could have employed a lighter touch but cheerfulness
prevails.
Thorn EMI VHS, Beta A

Forbidden Planet*
US 1956 98m ☐ colour

Walter Pidgeon, Anne Francis, Leslie
Nielsen, Warren Stevens
D, Fred M Wilcox

Ingenious attempt to find a sci-fi equivalent
of Shakespeare's *The Tempest* with Pidgeon
as a Prospero-figure conjuring up monsters;
Anne Francis in the Miranda role and
Robby the Robot standing in for Caliban.
MGM/UA VHS, Beta B

Futureworld
US 1976 104m v colour

Peter Fonda, Blythe Danner, Arthur Hill,
Yul Brynner, John Ryan
D, Richard T Heffron

Sequel to *Westworld* (qv), with a holiday centre in the grip of a mad scientist bidding to rule the world with robot clones of leading statesmen. An intriguing idea, which director Heffron plays mainly for fun.
Guild VHS, Beta, V2000 A

Invasion of the Body Snatchers*
US 1978 144m ● colour

Donald Sutherland, Brooke Adams,
Leonard Nimoy, Veronica Cartwright
D, Philip Kaufman

The pod people are taking over San Francisco. This re-make of a classic from the 1950s acknowledges the debt with parts for that film's leading man (Kevin McCarthy) and director (Don Siegel).
Warner VHS, Beta A

It Came from Outer Space*
US 1953 77m v bw

Richard Carlson, Barbara Rush, Charles
Drake, Kathleen Hughes
D, Jack Arnold

A groggy spacecraft lands in the Arizona desert and is repaired by its occupants assuming human identities. Loosely derived from a story by Ray Bradbury and shot originally in 3D, it makes good use of stark locations.
CIC VHS, Beta B

The Land That Time Forgot*
GB 1974 86m □ colour

Doug McClure, John McEnery, Susan
Penhaligon, Keith Barron
D, Kevin Connor

Tale by Tarzan-creator Edgar Rice Burroughs about the survivors of a First World War sea skirmish who find themselves on an island with prehistoric monsters. Sci-fi writer Michael Moorcock helped with this lively screen version.
Thorn EMI VHS, Beta A

Logan's Run
US 1976 115m v colour

Michael York, Richard Jordan, Jenny
Agutter, Roscoe Lee Browne
D, Michael Anderson

In the year 2274 people live an idyllic existence but are "terminated" at the age of 30; York and Agutter try to escape. Promising theme, weakly developed, which surprisingly won an Oscar for special effects.
MGM/UA VHS, Beta, V2000 A

The Andromeda Strain

Close Encounters of the Third Kind

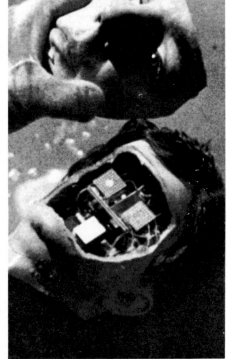

Futureworld

Lord of the Rings
US 1978 127m ☐ colour
Voices: Christopher Guard, John Hurt,
William Squire, Michael Scholes
D, Ralph Bakshi

Cartoon version of the Tolkien fantasy about Frodo the Hobbit and his mission to cast a magic ring back into its melting pot. An interesting departure by the director is that the characters are based on live-action photographs.
Thorn EMI VHS, Beta A

The Man Who Fell to Earth*
GB 1976 134m ● colour
David Bowie, Rip Torn, Candy Clark,
Buck Henry, Bernie Casey
D, Nicolas Roeg

Dense, obscure, tantalising film starring a wan David Bowie as the mysterious visitor from another planet with a complex plan to secure water for his dying family. Director Roeg's talent is beyond question, but here it submerges him.
Thorn EMI VHS, Beta A

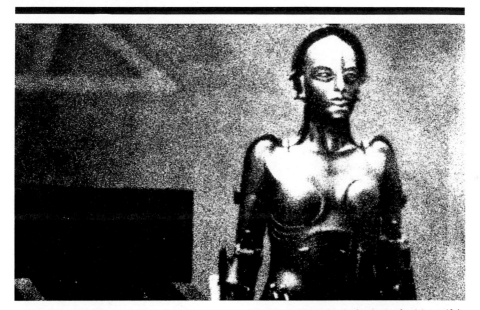

Metropolis***
Germany 1926 86m v bw
Alfred Abel, Gustav Fröhlich, Brigitte
Helm, Rudolf Klein-Rogge, Fritz Rasp
D, Fritz Lang

As late as the 1950s Fritz Lang's vision of a city of the future where the mass of people are enslaved by a small ruling elite was quoted as almost the only serious contribution that the cinema had made to the science fiction genre. Certainly there was nothing in the silent era that matched its epic sweep or the power of its imagination. It was one of the most expensive films of its period, took 310 days and 60 nights to shoot and used a cast of more than 700 as well as more than 30,000 extras. Written by Lang and his then wife, Thea von Harbou, the story of *Metropolis* is a revolt by the slaves against their masters, deliberately fomented by a mad scientist so that he can gain control of the city for himself. In a narrow sense, it is the battle between capital and labour; in a wider context (and this was Lang's own interpretation) it is about the benefits of modern science set against the magic of the dark ages. One of the disappointments of the film is that Lang and Von Harbou resolve the conflict so flatly, by asserting that love will win out in the end. It must also be admitted that the characters are no more than types and that the acting is less than subtle (Brigitte Helm, incidentally, plays the first cinema robot). What matters, though, is the grand design: the extraordinary, studio-constructed evocation of a sky-scrapered city (inspired, it is said, by a visit Lang made to New York); the dramatic lighting; the masterly handling of crowd scenes. Even in the video version, rather shorter than the usual two hour running time, the magic still comes across.
Thorn EMI VHS, Beta A

225

Plan 9 from Outer Space*
US 1956 79m □ bw

Bela Lugosi, Tor Johnson, Lyle Talbot,
Vampire, Gregory Walcott
D, Edward D Wood Jnr

Lugosi, in his last film, stalks a San Francisco graveyard. Unfortunately he died after completing only five minutes footage and was replaced by a stand-in who looked nothing like him. Dubbed "the worst film of all time". Unintentionally hilarious.
Palace   VHS, Beta   C

Planet of the Apes**
US 1968 119m v colour

Charlton Heston, Roddy McDowall, Kim
Hunter, Maurice Evans, James Whitmore
D, Franklin Schaffner

Astronaut Heston and friends crash-land on an unknown planet ruled by English-speaking apes. Intelligent fable about the regression of civilisation, with a stunning use of landscape; spawned many sequels.
CBS/Fox   VHS, Beta, Laser   A

Quintet
US 1979 113m v colour

Paul Newman, Vittorio Gassman,
Fernando Rey, Bibi Andersson
D, Robert Altman

Blasted by the critics on release, this may be due for reassessment. A sombre, enigmatic study of people playing games of death in an icebound city of the future. Pretentious or profound? – only time will tell.
CBS/Fox   VHS, Beta   A

Rollerball
US 1975 123m ● colour

James Caan, John Houseman, Ralph
Richardson, Maud Adams, John Beck
D, Norman Jewison

Heavy-handed parable of the future. A society no longer afflicted with poverty, sickness and war vents anti-social feelings through a violent game. Richardson's eccentric librarian provides welcome light relief.
Warner   VHS, Beta, Laser, CED   A

Silent Running*
US 1971 86m v colour

Bruce Dern, Cliff Potts, Ron Rifkin, Jesse
Vint
D, Douglas Trumbull

Sci-fi with an ecological message, charting astronaut Dern's obsessive determination to keep alive the one remaining example of earthly vegetation. An uneven film but with superb sequences, striking images and ultimately very moving.
CIC   VHS, Beta   B

Soylent Green*

US 1973 94m v colour

*Charlton Heston, Edward G Robinson,
Leigh Taylor-Young, Chuck Connors
D, Richard Fleischer*

A terrible warning about over-population with 40 million New Yorkers unwittingly having to exist on unsavoury synthetic food. Heston discovers the horrible secret after a deathbed tip-off by Edward G, in his last film.

MGM/UA VHS, Beta A

Star Trek – The Motion Picture

US 1979 130m □ colour

*William Shatner, Leonard Nimoy,
DeForest Kelley, Stephen Collins
D, Robert Wise*

Admiral Kirk resumes command of star ship Enterprise, as a huge malignant force heads for earth. Belated spin-off from TV, but money and a big screen are not everything.

CIC VHS, Beta, V2000, Laser A

Star Wars**

US 1977 116m □ colour

*Mark Hamill, Harrison Ford, Carrie
Fisher, Peter Cushing, Alec Guinness
D, George Lucas*

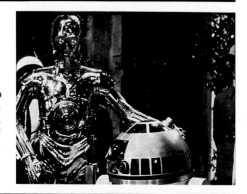

Power-crazed Peter Cushing attempts to take over the universe; clean-limbed heroes Hamill and Ford and beautiful princess Fisher stand in the way. Potent mixture of comic strip heroics and dazzling hardware.

CBS/Fox VHS, Beta, V2000, Laser A

Superman*

GB 1978 129m □ colour

*Christopher Reeve, Marlon Brando, Gene
Hackman, Margot Kidder, Jackie Cooper
D, Richard Donner*

The red-caped hero from the planet Krypton pits his wits and powers against master criminal Gene Hackman, who wants to blow California into the sea. Brando got $3.7m for his 10-minute spot; Superman Reeve gives better value.

Warner VHS, Beta A

Superman II*
US 1980 127m v colour

Christopher Reeve, Gene Hackman, Ned Beatty, Jackie Cooper, Sarah Douglas
D, Richard Lester

Mad general Zod is terrorising the United States but Superman has got married and lost his super-powers. The change of director (Lester, who made the Beatles films) makes this sequel a livelier, jokier affair.
Warner   VHS, Beta   A

Superman III*
US 1982 120m v colour

Christopher Reeve, Richard Pryor, Robert Vaughn, Pamela Stephenson, Annette O'Toole
D, Richard Lester

This time he is up against computer wizard Richard Pryor, who is programming a weather satellite so that it will destroy the Columbian coffee crop. Director Lester keeps it bubbling.
Thorn EMI   VHS, Beta   A

The Thing from Another World*
US 1951 83m v bw

Robert Cornthwaite, Kenneth Tobey, Margaret Sheridan, Bill Self
D, Christian Nyby

American scientists at the North Pole are menaced by a monstrous alien. Cogent mix of sci-fi and horror which has become a cult. With: *Stranger on the Third Floor*, 1940 drama.
Kingston   VHS, Beta   A

Things to Come**
GB 1936 108m v bw

Raymond Massey, Edward Chapman, Ralph Richardson, Margaretta Scott
D, William Cameron Menzies

H G Wells scripted his own vision of a world destroying itself by war. William Cameron Menzies came from the United States to mastermind the striking sets; and Arthur Bliss supplied a memorable score.
Spectrum   VHS, Beta   A

THX 1138*
US 1970 92m ● colour

Robert Duvall, Donald Pleasence, Pedro Colley, Maggie McOmie, Ian Wolfe
D, George Lucas

A hint of Orwell's *Nineteen Eighty-Four* in this cryptic directorial debut by George (*Star Wars*) Lucas. It depicts a computer-programmed society of the future in which people are shaven-headed automatons known only by numbers.
Warner   VHS, Beta   A

Time After Time*

US 1979 108m v colour

Malcolm McDowell, David Warner, Mary Steenburger, Charles Ciofi
D, Nicholas Meyer

McDowell as H G Wells, using his time machine to follow Jack the Ripper to present-day San Francisco. Ingenious fantasy, which makes some amusing points, better served by Meyer the writer than Meyer the director.
Warner VHS, Beta A

Tron*

US 1982 96m v colour

Bruce Boxleitner, Jeff Bridges, David Warner, Barnard Hughes
D, Steven Lisberger

Computer ace Bridges plays a video game to the death to eliminate his unscrupulous adversary; imaginative creation of the electronic world, mixing live action with computer graphics, with a welcome touch of humour.
Disney VHS, Beta R

20,000 Leagues Under the Sea*

US 1954 122m □ colour

Kirk Douglas, James Mason, Paul Lukas, Peter Lorre, Robert J Wilke
D, Richard Fleischer

The sinister Captain Nemo (Mason) captures Douglas and friends in his atomic-powered submarine *Nautilus*. Lively rendering of Jules Verne's classic Victorian adventure, with good tricks.
Disney VHS, Beta R

Quintet

Superman III

2001: A Space Odyssey***
GB 1968 134m v colour

*Keir Dullea, Gary Lockwood, William
Sylvester, Douglas Rain
D, Stanley Kubrick*

Having made *Dr Strangelove,* a macabre
film about nuclear war, it was logical that
for his next picture Stanley Kubrick should
go further into the future and tackle the
space age. His inspiration was a short story
by Arthur C Clarke called *The Sentinel,*
which had been published in 1951; Clarke
collaborated with Kubrick on the screenplay. In this case, however, screenplay
should not necessarily be taken as meaning words for much of the film has no
dialogue (it is nearly 30 minutes before the
first word is uttered). Kubrick called *2001* a
"non-verbal experience", largely dispen-
sing with conventional plotting as well as
inviting the spectator to focus on ideas
suggested by the images. The central
image is a mysterious black column which
turns up in the prologue — set in prehistoric times — and then, after the film
has jumped four million years in time, on
the moon, giving out radio signals which
send a group of astronauts towards Ju-
piter. What all this means — man trying to
push out the barriers of knowledge is the
easy summary — is something to ponder
while absorbing Kubrick's qualities as a
film maker. This is partly a matter of the
special effects, personally supervised by
him and among the most brilliant attemp-
ted in the cinema at that time; but also
some startling play with light and colour. Its
critics find *2001* slow and obscure; there is
something to both charges. It is also one of
the cinema's most ambitious attempts to
escape from the usual mechanics to plot
and dialogue and appeal to the audience's
imagination.

MGM/UA VHS, Beta, V2000, Laser A

Voyage to the Bottom of the Sea

US   1961   101m   □   colour

*Walter Pidgeon, Robert Sterling, Joan
Fontaine, Peter Lorre, Barbara Eden
D, Irwin Allen*

Atomic submarine commander Pidgeon on
his way to explode a deadly radiation belt
with a crew determined to cause trouble.
Agreeable piece of nonsense, ideal for
youngsters; it led to a popular TV series.
CBS/Fox   VHS, Beta   A

The War of the Worlds*

US   1952   85m   ●   colour

*Gene Barry, Ann Robinson, Les
Tremayne, Robert Cornthwaite, Sandra
Giglio
D, Byron Haskin*

Martian war machines invade Los Angeles
in a version of the Wells story updated to
include the atomic bomb. The hissing
attackers and a briefly glimpsed monster
are more convincing than the humans.
CIC   VHS, Beta   A

Westworld*

US   1973   87m   v   colour

*Yul Brynner, Richard Benjamin, James
Brolin, Norman Bartold
D, Michael Crichton*

Brynner, black-clad and eyes aglow, as a
robot gunslinger running amok in a rich
people's holiday centre. Writer/director
Crichton's dialogue is banal, but the
excitement compensates.
MGM/UA   VHS, Beta, CED   A

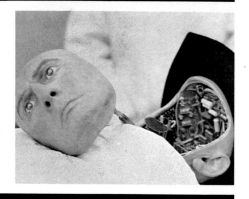

Zardoz

GB   1973   102m   ●   colour

*Sean Connery, Charlotte Rampling, John
Alderton, Sara Kestelman
D, John Boorman*

Complex and intellectual piece, set 300
years after the collapse of industrial society
when the earth has become a polluted
wasteland and a commune of scientists
holds the key to knowledge.
CBS/Fox   VHS, Beta, Laser   A

Zero Population Growth

US   1971   95m   ●   colour

*Oliver Reed, Geraldine Chaplin, Diane
Cilento, Don Gordon, Bill Nagy
D, Michael Campus*

In the twenty-first century there are so
many people that the world government
decides no more children shall be born for
30 years; young marrieds Reed and Chaplin
defy the ban. Bright idea, sentimentally
treated.
World Wide   VHS, Beta, V2000   A

AGATHA CHRISTIE'S EVIL UNDER THE SUN

A JOHN BRABOURNE/RICHARD GOODWIN Production

Starring

PETER USTINOV, JANE BIRKIN

COLIN BLAKELY, NICHOLAS CLAY

JAMES MASON, RODDY McDOWALL

SYLVIA MILES, DENIS QUILLEY

DIANA RIGG, MAGGIE SMITH

The American Friend*
WG/France 1977 123m v colour

Dennis Hopper, Bruno Ganz, Gérard Blain, Lisa Kreuzer
D, Wim Wenders

Hopper as Patricia Highsmith's Tom Ripley, finding dirty work for friend Ganz who turns hired killer to support his family. Absorbing interplay of character, two cleverly staged murders and loads of cynicism.
Palace VHS, Beta A

The Anderson Tapes*
US 1971 94m v colour

Sean Connery, Martin Balsam, Dyan Cannon, Alan King, Ralph Meeker
D, Sydney Lumet

New slant on the caper movie with ex-con Connery and friends planning to burgle a plush New York apartment, not knowing that every one of them is being bugged or under surveillance. Director Lumet keeps it crisp.
RCA/Columbia VHS, Beta A

Angel*
Ireland 1982 90m ● colour

Stephen Rea, Veronica Quilligan, Alan Devlin, Peter Caffrey
D, Neil Jordan

Promising first film by writer/director Jordan, set against the troubles in Ireland, with a young saxophonist (Rea) hunting down the gunmen who have killed a mute girl. Gripping study of obsessive vengeance.
Palace VHS, Beta A

Assault on Precinct 13**
US 1976 91m ● colour

Austin Stoker, Darwin Joston, Laurie Zimmer, Martin West, Nancy Loomis
D, John Carpenter

Director Carpenter showing he knows his Howard Hawks with this homage to *Rio Bravo* (qv) in which young hoodlums lay siege to a Los Angeles police station. Tense, gory, cleverly paced story of people under stress.
Media VHS, Beta A

Berlin Express*
US 1948 83m v bw

Merle Oberon, Robert Ryan, Charles Korvin, Paul Lukas, Robert Coote
D, Jacques Tourneur

Influential German statesman Lukas kidnapped by neo-Nazis in post-war Frankfurt; Allies Ryan and Oberon join the search. Effective Cold War thriller, with good performances. With: *Isle of the Dead* (qv).
Kingston   VHS, Beta   A

The Big Sleep***
US 1946 114 v bw

Humphrey Bogart, Lauren Bacall, Martha Vickers, John Ridgely, Dorothy Malone
D, Howard Hawks

The classic screen version of Raymond Chandler's downbeat thriller, full of crackling wit and moody atmosphere; Bogart as private eye Philip Marlowe. Don't worry about the plot – director Hawks couldn't understand it either!
Warner   CED only   C

The Big Steal*
US 1949 68m v bw

Robert Mitchum, Jane Greer, William Bendix, Ramon Novarro, Patric Knowles
D, Don Siegel

Taut and vigorous little film, which shows what can be done on a small budget. Mitchum plays an army officer framed for a payroll theft in hot pursuit of Bendix who set him up. With: *Underwater!* 1955 adventure.
Kingston   VHS, Beta   A

The Bird with the Crystal Plumage*
Italy/WG 1969 94m ● colour

Tony Musante, Suzy Kendall, Enrico Maria Salerno, Eva Renzi
D, Dario Argento

American writer Musante sees a man in black stabbing a woman in an art gallery; or is he imagining things? Learning a trick or two from Hitchcock, director Argento manages some neat twists before the end.
Videomedia   VHS, Beta, V2000   A

Blackmail**
GB 1929 90m v bw

Anny Ondra, Sara Allgood, John Longden, Charles Paton, Donald Calthrop
D, Alfred Hitchcock

Renowned as the first British talkie and showing Hitchcock quickly at home in the new medium. The breadknife sequence is the most famous but there is also a memorable chase through the British Museum; primitive but delightful.
Thorn EMI   VHS, Beta   C

Blow Out

US 1981 108m ● colour

John Travolta, Nancy Allen, John Lithgow, Dennis Franz
D, Brian de Palma

Sound effects man Travolta smells a cover-up when a politician is killed in a car crash; but he has the incident on tape and film and is determined to find the truth. Slick thriller with echoes of *The Conversation* (qv).
Rank VHS, Beta, Laser, CED A

Body Heat*

US 1981 113m ● colour

William Hurt, Kathleen Turner, Richard Crenna, Ted Danson
D, Lawrence Kasdan

Sexy, seductive up-date of that great 1940s film noir, *Double Indemnity,* with hard-up lawyer Hurt and wealthy mistress Turner in a plot to kill her husband. The original is still better but it's a very good try.
Warner VHS, Beta A

Bonnie and Clyde***

US 1967 109m ● colour

Warren Beatty, Faye Dunaway, Gene Hackman, Estelle Parsons, Michael J Pollard
D, Arthur Penn

Bonnie Parker and Clyde Barrow robbed banks during the American Depression, killed 18 people in the process and met an inevitable end in a hail of bullets. Arthur Penn's movie is one of several made about them or based on their exploits. It might have gone to France because the script was offered to François Truffaut and to Jean-Luc Godard before Warren Beatty persuaded Warner Brothers to back the project with himself as producer and leading man. The result was not only a big commercial success, which revived Beatty's erratic career and made immediate stars out of Dunaway and Hackman, but also one of the finest contributions to the gangster genre. Penn obeys the traditional Hollywood requirement of glamorous characters with whom the audience can identify (the real Bonnie and Clyde were nothing like as good looking as Dunaway and Beatty); he evokes a cosy nostalgia for the small-town America of the 1930s and, at the same time, he pulls the spectator up short with the graphic depiction of violence and its – literally – bloody consequences. So there is a constant tension between the desire to side with the fugitives and the realisation that their activities are contemptuous of human life. It is a rich, disturbing work and after it the gangster film was never to be the same again.
Warner VHS, Beta A

Borsalino & Co

Fra/Italy/WG 1974 107m ● col

Alain Delon, Catherine Rouvel, Riccardo Cucciolla, René Kolldehoff
D, Jacques Deray

Sequel to the 1970 film, *Borsalino*, continuing the saga of gangsters in pre-war Marseilles. Avenging the murder of his chum in the first film, Delon finds himself in a lot more trouble but has the last laugh.
VTC VHS, Beta, V2000 A

The Boston Strangler*

US 1968 110m ● colour

Henry Fonda, Tony Curtis, George Kennedy, Sally Kellerman
D, Richard Fleischer

Solid, low-key account of the real case of the 1960s mass murderer, played by Tony Curtis; Fonda as the attorney in charge of the investigation, Kellerman the only victim who survives to tell the tale.
CBS/Fox VHS, Beta, Laser A

The Boys from Brazil

US/GB 1978 120m ● colour

Gregory Peck, Laurence Olivier, James Mason, Lilli Palmer,
D, Franklin Schaffner

A German war criminal (Peck) tries to revive the master race by cloning a new breed of little Hitlers; Olivier as a Nazi hunter who gets wise to the plan. Far-fetched to start with and Schaffner's erratic direction does not help.
Precision VHS, Beta, V2000, CED A

Brighton Rock*

GB 1947 89m v bw

Richard Attenborough, Hermione Baddeley, Harcourt Williams, William Hartnell
D, John Boulting

Filleted version of Graham Greene's famous novel of Brighton gangland. It scores on location and atmosphere, with a compelling performance by Attenborough as the teenage hoodlum who kills his rival.
Thorn EMI VHS, Beta C

Bullitt*

US 1968 109m v colour

Steve McQueen, Jacqueline Bisset, Robert Vaughn, Don Gordon
D, Peter Yates

McQueen as the honest cop guarding a key witness; Vaughn as the attorney who is on his back; crisp work by director Yates. McQueen did his own driving in the pulsating 11-minute car chase through the streets of San Francisco.
Warner VHS, Beta A

The Butcher (Le Boucher)**
France/Italy 1969 92m ● col

*Stéphane Audran, Jean Yanne, Antonio
Passalia, Mario Beccaria
D, Claude Chabrol*

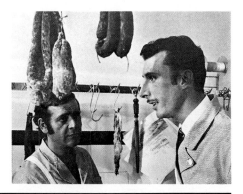

Thriller with the Hitchcock touch. A placid
French village is shocked by a series of
brutal murders and suspicion falls on the
kindly butcher. Masterly handling by
Chabrol of character and locale.
VCL VHS, Beta A

Chinatown**
US 1974 131m ● colour

*Jack Nicholson, Faye Dunaway, John
Huston, Perry Lopez, John Hillerman
D, Roman Polanski*

Scintillating latter-day excursion into Raymond Chandler territory. A matrimonial
private eye (Nicholson) uncovers murder
and corruption in 1937 Los Angeles; with
Dunaway as the double-dealing femme
fatale.
CIC VHS, Beta, Laser A

Coma*
US 1977 104m v colour

*Geneviève Bujold, Michael Douglas,
Richard Widmark, Elizabeth Ashley
D, Michael Crichton*

Ghoulish goings-on at a Boston hospital,
where patients are being bumped off so
that their bodies can be used for spare part
surgery. Bujold plays the young surgeon
who suspects; writer-director Crichton
keeps it tense.
MGM/UA VHS, Beta, V2000, Laser A

Cornered*
US 1945 99m v bw

*Dick Powell, Micheline Cheirel, Walter
Slezak, Morris Carnovsky, Nina Vale
D, Edward Dmytryk*

Crisp, unpretentious revenge thriller which
gets straight to the point and stays there.
Powell as a Canadian airman hunting
down his wife's killer (Slezak) in South
America. With: *The Woman on Pier 13*,
1949 drama.
Kingston VHS, Beta A

Blackmail

Blow Out

Brighton Rock

Crackup*
US 1946 90m v bw

Pat O'Brien, Claire Trevor, Herbert Marshall, Ray Collins
D, Irving Reis

Art expert Pat O'Brien loses his memory and is manipulated by a gang of forgers. Entertaining and unusual mystery with a particularly fine start. With: *I Walked With a Zombie* (qv).
Kingston   VHS, Beta   A

Cry Danger*
US 1951 75m v bw

Dick Powell, Rhonda Fleming, Richard Erdman, William Conrad
D, Robert Parrish

Released on parole after serving five years in jail, convicted robber Powell fights to clear his name. Fast moving, tautly directed melodrama which is a model of its type. With: *Run of the Arrow* (qv).
Kingston   VHS, Beta   A

Cutter's Way*
US 1981 106m v colour

Jeff Bridges, John Heard, Lisa Eichhorn, Ann Dusenberry
D, Ivan Passer

Cutter (John Heard) is an embittered Vietnam war veteran who decides to blackmail an oil tycoon he suspects of committing a murder; friend Jeff Bridges is dragged in. Dense, enigmatic piece, full of teasing questions.
Warner,   VHS, Beta   A

The Day of the Jackal*
GB/France 1973 136m v colour

Edward Fox, Alan Badel, Eric Porter, Cyril Cusack, Delphine Seyrig
D, Fred Zinnemann

Careful, craftsmanlike adaptation of the Frederick Forsyth best-seller about a French generals' plot to assassinate President de Gaulle. Fox is impressive as the hired killer but plot mainly triumphs over character.
CIC   VHS, Beta   B

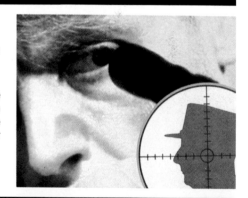

Dead Pigeon on Beethoven Street*
WG 1972 96m v colour

Glenn Corbett, Christa Lang, Anton Diffring, Eric P Caspar
D, Samuel Fuller

Extended in-joke by cult director Fuller; full of references to other movies, and that will appeal to buffs. Corbett as an American private eye in Bonn, on the trail of a hit man.
Thorn EMI   VHS, Beta   A

Deadline at Dawn*
US   1946   80m   v   bw

*Paul Lukas, Bill Williams, Susan
Hayward, Osa Massen, Lola Lane*
D, Harold Clurman

A rare excursion into the cinema by theatre director Clurman. This is an intriguing piece about a young sailor on leave in New York who finds himself implicated in a murder. With *Stagecoach* (qv).
Kingston   VHS, Beta   A

Death Wish
US   1974   93m   ●   colour

*Charles Bronson, Hope Lange, Vincent
Gardenia, Stuart Margolin*
D. Michael Winner

Thick-eared vigilante film, directed by Winner in his hectic sledgehammer style. Bronson portrays a New York businessman seeking personal revenge after his wife and daughter are assaulted by muggers.
CIC   VHS, Beta, V2000, Laser   A

Deathtrap
US   1982   113m   ●   colour

*Michael Caine, Christopher Reeve, Dyan
Cannon, Irene Worth*
D, Sidney Lumet

Serviceable adaptation of Ira Levin's ingeniously plotted stage hit about a playwright (Caine) who has a string of flops and decides to murder a young protegé (Reeve) and steal his latest work. But all is not what it seems.
Warner   VHS, Beta   A

The Deep
US   1977   123m   ●   colour

*Jacqueline Bisset, Robert Shaw, Nick
Nolte, Lou Gossett, Eli Wallach*
D, Peter Yates

Bisset and Nolte diving in the waters off Bermuda stumble on a consignment of morphine sunk during the Second World War. Voodoo and a monster eel add spice to a thin meal; director Yates supplies his usual technical polish.
RCA/Columbia   VHS, Beta, CED   A

Dial M for Murder**
US   1954   96m   v   colour

*Ray Milland, John Williams, Grace Kelly,
Robert Cummings*
D, Alfred Hitchcock

Tennis star Milland hatches elaborate plot to dispose of wife Kelly to get his hands on her money. Hitchcock shows how to make compelling cinema out of a one-set play which relies almost entirely on dialogue.
VTC   VHS, Beta   A

Diamonds Are Forever*
GB 1971 119m v colour
Sean Connery, Jill St John, Charles Gray, Lana Wood, Jimmy Dean
D, Guy Hamilton
The villain Blofeld is plotting to blackmail Washington with a giant laser and James Bond has to stop him. Above-average entry in the 007 stakes, with Connery demonstrating how much his laconic style was later missed.
Warner VHS, Beta A

Dirty Harry*
US 1971 99m ● colour
Clint Eastwood, Harry Guardiano, Reni Santoni, John Vernon
D, Don Siegel
Eastwood as the violent San Francisco cop going against the orders of his superiors to nail a psychopathic killer. Set off a crop of "taking the law into your own hands" movies, few of them made with Siegel's cool skill.
Warner VHS, Beta A

Diva**
France 1981 120m ● colour
Frederic Andrei, Roland Bertin, Richard Bohringer, Wilhelmenia Wiggins Fernandez
D, Jean-Jacques Beineix
Original, stylish concoction with some well-aimed digs at the thriller formula. A black American opera singer is secretly recorded by a young admirer who finds himself caught up in drugs and murder.
Palace VHS, Beta A

Dr Mabuse – The Gambler *
Germany 1922 95m v bw
Rudolph Klein-Rogge, Alfred Abel, Gertrude Welcker, Lil Dagover
D, Fritz Lang
One of the classics of early German cinema, it is about a criminal bent on world control. Ponderous and confusing; but a film of great power, clearly reflecting post-war chaos.
Thorn EMI VHS, Beta C

Doctor No*
GB 1962 111m v colour
Sean Connery, Ursula Andress, Jack Lord, Joseph Wiseman
D, Terence Young
The first of the James Bond films, it established a genre that was still being milked more than 20 years later. Sean Connery as suave agent 007, up against master criminal Wiseman in Jamaica; Andress in decorative support.
Warner VHS, Beta A

Dressed to Kill*
US 1980 104m ● colour
Michael Caine, Angie Dickinson, Nancy Allen, Keith Gordon, Dennis Franz
D, Brian De Palma
Flashy, adroit, unnerving shocker about a sexually frustrated wife (Dickinson) who goes to psychiatrist Caine with her troubles and meets a horrible end in a lift. *Psycho* (qv) revisited but without the Hitchcock magic.
Guild VHS, Beta, V2000, CED A

Evil Under the Sun
GB 1981 112m v colour

Peter Ustinov, James Mason, Diana Rigg,
Maggie Smith, Colin Blakely
D, Guy Hamilton

The stars line up for a crack at Agatha Christie, with Ustinov as sleuth Hercule Poirot exercising his little grey cells on an idyllic Adriatic island. Spirited acting and pretty locations compensate for laborious plotting.
Thorn EMI   VHS, Beta   A

Executive Action
US 1973 91m v colour

Burt Lancaster, Robert Ryan, Will Geer,
Gilbert Green, John Anderson
D, David Miller

Enactment of the theory that the assassination of President Kennedy was plotted by a right-wing group (headed here by Lancaster and Ryan) using Lee Harvey Oswald as a fall guy. The fiction lies uneasily with the fact.
Warner   VHS, Beta   A

The Eyes of Laura Mars
US 1978 103m v colour

Faye Dunaway, Tommy Lee Jones, Brad Dourif, René Auberjonois
D, Irvin Kershner

Fashion photographer Faye Dunaway has premonitions about a series of murders; detective Tommy Lee Jones investigates. Promising idea swamped by colour-supplement trendiness but it has a certain eye-catching chic.
RCA/Columbia   VHS, Beta, CED   A

The Face of Fu Manchu*
GB 1965 94m □ colour

Nigel Green, Christopher Lee, Tsai Chin, Joachim Fuchsberger
D, Don Sharp

Sax Rohmer's oriental master criminal, Dr Fu Manchu (Lee) is up to his tricks again, trying to dominate the world with a killer gas. But Nayland Smith of Scotland Yard (Green) is on the scent. Stylish, unpatronising fun.
Thorn EMI   VHS, Beta   A

Family Plot*
US 1976 115m v colour

Karen Black, Bruce Dern, Barbara Harris, William Devane, Ed Lauter
D, Alfred Hitchcock

A fake medium, a professional kidnapper, a long-lost child, a bishop abducted from his cathedral; Alfred Hitchcock's last film is a delicious confection of plot and counterplot, with the old boy in playful mood to the end.
CIC   VHS, Beta   A

Farewell My Lovely*
US 1975 92m v colour

Robert Mitchum, Charlotte Rampling, John Ireland, Sylvia Miles
D, Dick Richards

A worthy and affectionate re-make of a 1940s classic, true to atmosphere and period. Mitchum as Raymond Chandler's world-weary private eye, Philip Marlowe, on the trail of a stolen necklace and a missing girl.
Precision   VHS, Beta, CED   A

The FBI Story
US 1959 144m v colour

*James Stewart, Very Miles, Larry
Pennell, Nick Adams
D, Mervyn Le Roy*

Romantic, punch-pulling look at the American Federal Bureau of Investigation. It covers such episodes as the Ku Klux Klan, the 1930s gangsters and Nazi spy rings through an agent's reminiscences. Stewart makes it watchable.
Warner VHS, Beta A

Fear is the Key*
GB 1972 101m v colour

*Suzy Kendall, Barry Newman, John
Vernon, Dolph Sweet, Ben Kingsley
D, Michael Turner*

A clever Alistair MacLean plot sets Newman the task of discovering who shot down the plane carrying his wife and family. Uncomplicated and confidently handled film which starts with a car chase and never lets up.
Thorn EMI VHS, Beta A

Finally, Sunday!
France 1983 113m v bw

*Fanny Ardant, Jean-Louis Trintignant,
Phillip Laudenbach, Caroline Sihol
D, François Truffaut*

Estate agent Trintignant is the prime suspect when his wife and her lover are murdered; but loyal secretary (Ardant) sets out to clear his name. Lightweight comedy-thriller from a distinguished director marking time.
Thorn EMI VHS, Beta A

First Blood
US 1982 90m v colour

*Sylvester Stallone, Brian Dennehy,
Richard Crenna, David Caruso
D, Ted Kotcheff*

A Vietnam hero (Stallone) carries the scars of war into civilian life, beats up three policemen and goes on the run. Lurking within a sententious message about violence breeding violence is a gritty chase thriller.
Thorn EMI VHS, Beta A

The First Great Train Robbery
GB 1978 111m v colour

*Sean Connery, Donald Sutherland,
Lesley-Anne Down, Robert Lang
D, Michael Crichton*

Writer/director Crichton (*Westworld, Coma,* qqv) goes back to mid-Victorian England for a crafty caper about relieving the London Bridge to Folkestone train of its bullion. Connery and Sutherland go for gold.
Warner VHS, Beta A

For Your Eyes Only
GB 1981 127m v colour

*Roger Moore, Carole Bouquet, Topol,
Julian Glover, Jill Bennett
D, John Glen*

The James Bond saga rolls on, with Moore as 007 and Topol as the heavy. The plot is about a surveillance vessel sinking off the Greek coast and threatening to fall into the wrong hands. Saved by the stunts and trickery.
Warner VHS, Beta A

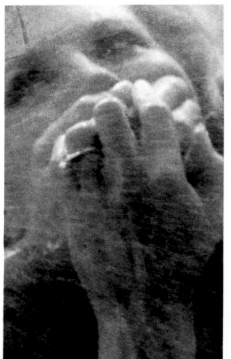

Dressed to Kill

Farewell My Lovely

First Blood

Fort Apache, the Bronx
US 1980 119m v colour
Paul Newman, Ed Asner, Ken Wahl,
Danny Aiello, Rachel Ticotin
D, Daniel Petrie

The everyday story of a decent, honest cop
(Newman), picking his way through the
social misfits of the South Bronx. A
contrived look at policing a tough neigh-
bourhood; Newman tries hard to make it
ring true.
VTC VHS, Beta A

48 Hours
US 1982 97m ● colour
Nick Nolte, Eddie Murphy, Annette
O'Toole, Frank McRae
D, Walter Hill

Nick Nolte and Eddie Murphy as white cop
and black convict form an uneasy alliance
to hunt down criminals and stolen money.
The violence is relieved by touches of
comedy and the personable playing of the
leading men.
CIC VHS, Beta, Laser A

The French Connection*
US 1971 100m ● colour
Gene Hackman, Roy Scheider, Fernando
Rey, Tony Lo Bianco
D, William Friedkin

Narcotics squad detectives Hackman and
Scheider smell out an international drugs
ring and do not mind whose toes they step
on. Convincing police procedural, based on
a real case, with a gripping New York car
chase.
CBS/Fox VHS, Beta, V2000, Laser A

French Connection II
US 1975 112m ● colour
Gene Hackman, Fernando Rey, Bernard
Fresson, Jean-Pierre Castaldi
D, John Frankenheimer

Hackman's New York cop back on the trail
of his old adversary, drugs pusher Fernando Rey, this time in Marseilles. Like
most sequels, this one misses the force of
the original but there are some good action
sequences.
CBS/Fox VHS, Beta, V2000, Laser A

Frenzy*
GB 1972 110m ● colour

Jon Finch, Alec McCowen, Barry Foster,
Vivien Merchant, Anna Massey
D, Alfred Hitchcock

A necktie murderer is terrorising London
and ex-RAF man Finch is prime suspect.
Leisurely, old fashioned Hitchcock film, let
down by a soggy script, but with enough of
the master's macabre touches to hold the
attention.
CIC VHS, Beta A

From Russia with Love**
GB 1963 118m v colour

Sean Connery, Robert Shaw, Lotte
Lenya, Daniela Bianchi
D, Terence Young

The second, and arguably the best, in the
James Bond cycle. Connery's cool cynicism
is nicely developed; with two formidable
villains in Shaw and Lenya and some
diverting set pieces that just stay within the
bounds of the credible.
Warner VHS, Beta A

The Getaway*
US 1972 123m ● colour

Steve McQueen, Ali MacGraw, Ben
Johnson, Sally Struthers, Al Lettieri
D, Sam Peckinpah

Ex-con McQueen shoots his way to the
Mexican border wth the loot from a bank
robbery; wife MacGraw gives dubious
support. Fast-moving chase thriller laced
with director Peckinpah's penchant for
mindless violence.
Warner VHS, Beta A

The Godfather***
US 1971 164m ● colour

Marlon Brando, Al Pacino, Robert Duvall,
James Caan, Sterling Hayden
D, Francis Ford Coppola

Long, absorbing saga of Mafia folk, with
Brandon croaking his way to an Oscar as
the family patriarch, Pacino the son who
takes over. Impressively assured film-
making, with a strong narrative grasp and
sharp performances.
CIC VHS, Beta A

The Godfather, Part II**
US 1974 192m ● colour

Al Pacino, Robert de Niro, Diane Keaton,
Robert Duvall, John Cazale
D, Francis Ford Coppola

Sequel which is only slightly less effective
than the original, tracing both the family's
Sicilian origins and the progress of Pacino's
reign. Robert De Niro plays the younger
version of the Brando character from Part I.
CIC VHS, Beta, V2000 A

Goldfinger**
GB 1964 108m v colour

Sean Connery, Honor Blackman, Gert
Frobe, Harold Sakata, Shirley Eaton
D, Guy Hamilton

James Bond foiling the dastardly Goldfinger's plan to blow up Fort Knox. One of
the most assured offerings of the whole
series, with good jokes, memorable villains
(Frobe's Goldfinger, Sakata's Odd-job) and
stunning sets.
Warner VHS, Beta A

The Grissom Gang
US 1971 128m ● colour

Scott Wilson, Kim Darby, Tony Musante,
Robert Lansing, Irene Dailey
D, Robert Aldrich

No holds barred re-make of *No Orchids For*
Miss Blandish with Kim Darby as the
kidnapped heiress who falls for her psychotic captor; director Aldrich gives it a camp
commercial gloss.
Guild VHS, Beta A

Hammett*
US 1982 93m v colour
Frederic Forrest, Peter Boyle, Marilu Henner, Roy Kinnear
D, Wim Wenders

Frederic Forrest as Dashiell Hammett, fulfilling his dual roles as private detective (called in by a colleague to find a missing Chinese girl) and crime writer. An accomplished work, belying its troubled production history.
3M Video VHS, Beta, V2000 A

Hell on Frisco Bay
US 1955 94m v colour
Alan Ladd, Edward G Robinson, Joanne Dru, Paul Stewart, William Demarest
D, Frank Tuttle

Alan Ladd as an ex-cop jailed for manslaughter and Edward G as the crook who framed him; exciting speedboat climax across San Francisco Bay. With: *The Mad Miss Manton,* 1938 comedy.
Kingston VHS, Beta A

High and Low*
Japan 1963 142m v colour
Toshiro Mifune, Kyoko Kagawa, Tatsuya Nakadai, Tatsuya Mihashi
D, Akira Kurosawa

Wealthy shoe manufacturer Mifune faces a huge ransom demand when his chauffeur's son is kidnapped in mistake for his own; to pay may mean financial ruin. Intriguing Japanese version of an American thriller by Ed McBain.
Palace VHS, Beta A

The Human Factor
GB 1979 115m v colour
Nicol Williamson, Richard Attenborough, Derek Jacobi, Robert Morley
D, Otto Preminger

Matter-of-fact adaptation by Tom Stoppard of Graham Greene's fine novel of spying and betrayal; it captures the plot but not the ambience. Williamson as the Foreign Office security man leading a double life.
Rank VHS, Beta A

I Confess*
US 1953 93m v bw
Montgomery Clift, Anne Baxter, Brian Aherne, Karl Malden
D, Alfred Hitchcock

The favourite Hitchcock theme of transfer of guilt given one its cleanest expressions in this tale of a Quebec priest (Clift) who hears a confession of murder but cannot divulge it and is himself accused of the crime.
Warner VHS, Beta A

For Your Eyes Only

From Russia with Love

Octopussy

In the Heat of the Night*
US 1967 109m v colour

*Sidney Poitier, Rod Steiger, Warren
Oates, Quentin Dean
D, Norman Jewison*

Bigoted southern sheriff (Steiger) and black
detective (Poitier) join forces on a murder
hunt; director Jewison using the thriller
format to make a plea for racial harmony,
with two stirring performances.
Warner VHS, Beta A

The Ipcress File*
GB 1965 108m v colour

*Michael Caine, Nigel Green, Guy
Doleman, Sue Lloyd, Gordon Jackson
D, Sidney J Furie*

Caine as Len Deighton's shabby, bespectacled secret agent (unnamed in the book,
here called Harry Palmer) on the trail of a
missing scientist. Low-key antidote to the
James Bond cycle, with the right seedy
atmosphere.
Rank VHS, Beta, Laser A

Jaws*
US 1975 124m v colour

*Robert Shaw, Roy Scheider, Richard
Dreyfuss, Lorraine Gary
D, Steven Spielberg*

Long Island holiday-makers are menaced
by a man-eating shark and Spielberg takes
the first step towards becoming a cinema
millionaire. Well calculated mixture of
thriller and disaster movie.
CIC VHS, Beta, LASER A
Also available: *Jaws II* (CIC)

Key Largo*
US 1948 101m v bw

*Humphrey Bogart, Lauren Bacall, Claire
Trevor, Edward G Robinson, Lionel
Barrymore
D, John Huston*

Gang boss Edward G Robinson holds
Bogart, Bacall and Barrymore hostage in a
Florida Keys hotel; static, talkative version
of Maxwell Anderson's play, but the
performances are too good to miss.
Warner VHS, Beta, CED A

The Killer Elite
US 1975 122m ● colour

*James Caan, Robert Duvall, Arthur Hill,
Gig Young, Mako, Bo Hopkins
D, Sam Peckinpah*

Study of shifting loyalties among members
of a private crime-fighting organisation in
San Francisco, with an injection of martial
arts. Routine stuff for director Peckinpah
but delivered with high professional gloss.
Warner VHS, Beta A

The Killers*
US 1964 91m ● colour

*John Cassavetes, Lee Marvin, Clu
Gulager, Angie Dickinson
D, Don Siegel*

Second screen version of the Hemingway
story about two hired assassins (icily
played by Marvin and Gulager) and their
mysterious assignment. Ronald Reagan
made his last screen appearance, before
going on to higher things.
CIC VHS, Beta B

Klute**
US 1971 109m ● colour

Jane Fonda, Donald Sutherland, Charles Ciofi, Roy Scheider, Rita Gam
D, Alan J Pakula

Jane Fonda in her Oscar-winning performance as a New York call girl, getting involved with private detective Sutherland in the search for a missing scientist. The plot comes second to a multi-layered study of voyeurism and identity.
Warner   VHS, Beta   A

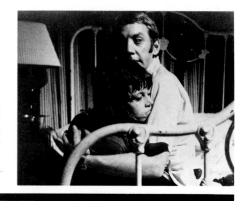

The Lady Vanishes***
GB 1938 93m v bw

Margaret Lockwood, Michael Redgrave, May Whitty, Cecil Parker, Basil Radford
D, Alfred Hitchcock

Dear old May Whitty disappears from a train in the middle of Europe; Lockwood and Redgrave turn sleuth to find out why. Fast-moving, witty and inventive Hitchcock spy thriller, his finest film before moving to the United States.
Rank   VHS, Beta, Laser   A

Last Embrace
US 1979 98m ● colour

Roy Scheider, Janet Margolin, Sam Leven, John Glover, Christopher Walken
D, Jonathan Demme

Scheider as a private investigator surviving several attempts on his life before he discovers what is behind them; Margolin the anthropology student with whom he gets involved. Conventional fare, presented with style.
Warner   VHS, Beta   A

The League of Gentlemen*
GB 1960 112m v bw

Jack Hawkins, Richard Attenborough, Roger Livesey, Nigel Patrick
D, Basil Dearden

Retired ex-army man Hawkins recruits a group of fellow officers to rob a London bank. Likeable and entertaining comedy-thriller, with a rich streak of British eccentricity and actors throughly enjoying themselves.
Rank   VHS, Beta   A

Lipstick
US 1976 86m ● colour

Margaux Hemingway, Perry King, Anne Bancroft, Chris Sarandon
D, Lamont Johnson

Fashion model is raped by a man who resents her beauty and success; but he is acquitted in court and strikes again. Cautionary tale about the consumer society, trying too hard to moralise; but the message is loud and clear.
CIC   VHS, Beta   A

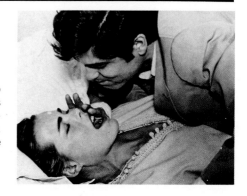

Little Caesar**
US 1930 77m ● bw

Edward G Robinson, Douglas Fairbanks Jnr, Glenda Farrell, William Collier Jnr
D, Mervyn Le Roy

Explosive early gangster classic, charting the rise and fall of a mobster based on Al Capone and played with demonic energy by Edward G Robinson. A bit rough at the edges but moves with tremendous pace and vitality.
Warner CED only C

Live and Let Die
GB 1973 121m v colour

Roger Moore, Yaphet Kotto, Jane Seymour, Clifton James, David Hedison
D, Guy Hamilton

Roger Moore makes his debut as James Bond, trying to smash a heroin racket in the Caribbean; Yaphet Kotto as the black master criminal. The cycle is settling into a formula, though the technology is still impressive.
Warner VHS, Beta A

The Long Good Friday**
GB 1980 109m ● colour

Bob Hoskins, Helen Mirren, Dave King, Bryan Marshall, Brian Hall
D, John Mackenzie

A sharp piece of movie making, although not quite as good as some of its admirers claim. Bob Hoskins impeccably cast as the London gang boss trying to hold his ground against a rival faction; Mirren as the moll.
Thorn EMI VHS, Beta, CED A

The Looking Glass War
GB 1969 103m v colour

Christopher Jones, Pia Degermark, Ralph Richardson, Anthony Hopkins
D, Frank R Pierson

Polish refugee (Jones) sent by British intelligence into East Germany to find film of a Russian missile site; on the way he finds glamorous Pia Degermark. Adequate version of the John Le Carré novel with good downbeat atmosphere.
RCA/Columbia VHS, Beta A

Loophole
GB 1980 105m v colour

Albert Finney, Martin Sheen, Susannah York, Colin Blakely
D, John Quested

Finney and chums wading through a sewer full of rats on their way to blasting open a London bank vault. Straightforward, old-fashioned heist which could just as well been made in 1950 as 1980; the cast is not overworked.
Brent Walker VHS, Beta A

McVicar*
GB 1980 98m ● colour

Roger Daltrey, Adam Faith, Cheryl Campbell, Billy Murray, Georgina Hale
D, Tom Clegg

Ex-con John McVicar co-scripted his own story of breaking out of Durham prison and living on the run; but he leaves the acting to Roger Daltrey. Brisk television-style thriller with a veneer of realism.
Spectrum VHS, Beta, V2000, Laser A

Macao
US 1952 78m v bw
Robert Mitchum, Jane Russell, William
Bendix, Gloria Grahame
D, Josef Von Sternberg

Moody Mitchum makes sexual chemistry
with Jane Russell in the Far East and helps
the police to nail a gangster. Interesting for
its director, though the film was completed
by Nicholas Ray. With: *The Bounty Hunter,*
1954 Western.
Kingston VHS, Beta A

Magnum Force
US 1973 115m ● colour
Clint Eastwood, Hal Holbrook, Mitch
Ryan, Felton Perry, David Soul
D, Ted Post

Clint Eastwood returns as the San Francisco detective, "Dirty" Harry Callahan,
meting out rough justice to a squad of
motorcycle cops even more vicious than he
is. The message is that the end justifies the
means.
Warner VHS, Beta A

The Maltese Falcon***
US 1941 91m v bw
Humphrey Bogart, Mary Astor, Sydney
Greenstreet, Peter Lorre, Elisha Cook Jnr
D, John Huston

The 1941 film of *The Maltese Falcon* was in
fact, the third made from Dashiell Hammett's novel – and who now remembers
the 1931 and 1936 versions, let alone the
actors who played the hard-boiled private
eye hero, Sam Spade (Ricardo Cortez and
Warren William)? Humphrey Bogart in
1941 was coming to his peak after years as
an indeterminate second or third lead and
The Maltese Falcon did as much as any film
to define the essential Bogart screen
character: tough, laconic, warm-hearted
and never fooled. (Ironically Warner
Brothers first offered the part to George Raft,
who turned it down because he did not want
to work with an untried director). But it is not
just Bogie's film. There is Mary Astor in her
finest screen role as the viperous Brigid
O'Shaughnessy; Peter Lorre, typically, as a
character on the edge of madness; the
huge Sydney Greenstreet on his film debut
at the ripe age of 62; and little Elisha Cook
Jnr, the cinema's eternal fall-guy. Huston
had been a screenwriter with Warner
Brothers during the 1930s and knew all
about shaping narrative and dialogue. Here
he showed how to get the best out of his
rich cast and a precise feel for the seedy,
corrupt ambience of Hammett's novel.
Huston was still going strong into the
1980s, but who can say that he has turned
out a more assured piece of work than this,
his first film?
Warner VHS, Beta A

The Man with the Golden Gun
GB   1974   125m   v   colour

Roger Moore, Christopher Lee, Britt Ekland, Maud Adams, Clifton James
D, *Guy Hamilton*

Christopher Lee takes time out from horror movies to play the suave baddie, Scaramanga, trying to nail James Bond with his golden bullets. Fair addition to the 007 saga, re-hashing all the familiar ingredients.
Warner   VHS, Beta   A

The Mean Machine*
US   1974   121m   ●   colour

Burt Reynolds, Eddie Albert, Ed Lauter, Michael Conrad, Jim Hampton
D, *Robert Aldrich*

Aldrich back in *The Dirty Dozen* (qv) groove. A football star (Reynolds) goes to jail to get up a team of convicts to play the guards. Vigorous, erratic and cheerfully cynical, with flashes of grotesque humour.
CIC   VHS, Beta   A

The Mechanic
US   1972   100m   ●   colour

Charles Bronson, Jan Michael Vincent, Keenan Wynn, Jill Ireland
D, *Michael Winner*

Bronson as a professional hit man playing cat and mouse with the son of one of his victims. Has its gripping moments but is mainly an excuse for a flashy director to show us his repertoire of cinematic gimmicks.
Warner   VHS, Beta   A

The Mirror Crack'd
GB   1980   100m   v   colour

Angela Lansbury, Geraldine Chaplin, Elizabeth Taylor, Rock Hudson, Tony Curtis
D, *Guy Hamilton*

Miss Marple (Angela Lansbury) is called in to investigate nasty goings-on in a quiet Kentish village. Another all-star assault on Agatha Christie, handsomely done but lacking her sense of humour.
Thorn EMI   VHS, Beta   A

Missing*
US   1982   116m   v   colour

Jack Lemmon, Sissy Spacek, Melanie Mayron, John Shea, Charles Cioffi
D, *Costa-Gavras*

A New York businessman (Lemmon) looks for his son who has disappeared in Chile during the military coup against Allende. Intricate political whodunit by the director of *Z*, skilfully evoking a country torn by strife.
CIC   VHS, Beta, V2000   A

Modesty Blaise
GB 1966 120m v colour

*Monica Vitti, Dirk Bogarde, Terence
Stamp, Harry Andrews
D, Joseph Losey*

Comic strip heroine Modesty (Vitti) trying to
save a shipment of diamonds from the
clutches of arch-villain Bogarde. Brightly
coloured hymn to the swinging sixties by a
"serious" director letting his hair down.
CBS/Fox VHS, Beta A

The Moon in the Gutter*
France/Italy 1983 126m ● colour

*Nastassia Kinski, Gérard Depardieu,
Victoria Abril, Vittorio Mezzogiorno
D, Jean-Jacques Beineix*

Moody study of a young doctor (Depardieu)
trapped by the memory of his dead sister
and determined to avenge himself on the
man who raped her and drove her to
suicide. Stylish piece of fatalism, with an
eye for the telling image.
Palace VHS, Beta A

Moonraker
GB 1979 121m v colour

*Roger Moore, Lois Chiles, Michael
Lonsdale, Richard Kiel, Geoffrey Keen
D, Lewis Gilbert*

A missing space shuttle, a megalomaniac
millionaire bent on world domination and a
seven-foot hit man: what more could a
James Bond movie want? Well, a bit of
inspiration would help, though the time
passes easily enough.
Warner VHS, Beta A

Murder on the Orient Express*
GB 1974 122m v colour

*Albert Finney, Ingrid Bergman, Lauren
Bacall, Vanessa Redgrave, John Gielgud
D, Sidney Lumet*

Ingenious Agatha Christie whodunit, set
aboard a snowbound train; Finney as
sleuth Hercule Poirot questioning an all-
star cast of suspects. Only one set and
nearly all talk but director Lumet manages
to hold the attention.
Thorn EMI VHS, Beta A

Never Say Never Again*
GB 1983 130m v colour

*Sean Connery, Max von Sydow, Klaus
Maria Brandauer, Barbara Carrera
D, Irvin Kershner*

Those SPECTRE nasties are trying to take
over the world again but Connery/Bond
has come out of retirement to thwart them.
A partial re-make of *Thunderball* (qv),
updated to take in holographic computers
but leaving the Connery style intact.
Warner VHS, Beta A

Nighthawks
US 1981 96m ● colour

*Sylvester Stallone, Billy Dee Williams,
Rutger Hauer, Lindsay Wagner
D, Bruce Malmuth*

Cop Stallone stalks an international ter-
rorist in New York. Efficiently made, television-style thriller presented as a showcase for its star; good locations; director
Malmuth cuts the cackle and gets on with
the action.
CIC VHS, Beta, V2000, Laser A

Night Moves*
US 1975 96m ● colour

Gene Hackman, Jennifer Warren,
Edward Binns, Harris Yulin
D, Arthur Penn

Private eye Hackman searches for a run-away girl in Florida and director Penn ensures that nothing is quite what it seems. Dense, enigmatic thriller slammed by the critics on first release but worth another look.
Warner   VHS, Beta   A

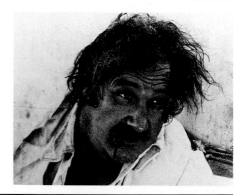

North by Northwest***
US 1959 136m v colour

Cary Grant, Eva Marie Saint, James
Mason, Leo G Carroll, Martin Landau
D, Alfred Hitchcock

Joyously tongue-in-cheek chase movie with Grant forced to go on the run after being mistaken for a spy. Magic moments include a murder at the United Nations; Grant being menaced by a crop-sprayer; and a shoot-out on Mount Rushmore.
MGM/UA   VHS, Beta   B

Notorious***
US 1946 100m v bw

Cary Grant, Ingrid Bergman, Claude
Rains, Louis Calhern
D, Alfred Hitchcock

A fast lady (Bergman) marries Nazi boss (Rains) in Rio, but only to expose him; an American agent (Grant) is the man she really loves. Expert spy thriller-cum-romance, with Hitchcock living up to his reputation as master of suspense.
Guild   VHS, Beta   A

Octopussy
GB 1983 127m v colour

Roger Moore, Louis Jourdan, Maud
Adams, Kristina Wayborn
D, John Glen

James Bond foiling a wicked commie plot to blow up a US air force base in West Germany. The usual concoction of gadgetry, girls and throwaway lines; but edges into the 1980s with references to nuclear disarmament.
Warner   VHS, Beta   A

On Her Majesty's Secret Service
GB 1969 130m v colour

George Lazenby, Diana Rigg, Telly
Savalas, Ilse Steppat
D, Peter Hunt

Lazenby's only and controversial appearance as James Bond. He tackles the blackguard Blofeld (Telly Savalas) who is trying to destroy the world with a deadly virus. Spectacular bobsleigh chase.
Warner   VHS, Beta   A

Out of the Past***
US 1947 93m v bw

*Kirk Douglas, Robert Mitchum, Jane
Greer, Rhonda Fleming, Richard Webb
D, Jacques Tourneur*

Outstanding example of the 1940s *film
noir*; private eye Mitchum comes out of
retirement to help hoodlum Douglas find
his girl (seductive Jane Greer) and becomes
ensnared by her. With: *Hell's Highway*,
1932 drama.
Kingston   VHS, Beta   A

The Paradine Case*
US 1947 116m v bw

*Gregory Peck, Ann Todd, Charles
Laughton, Alida Valli, Louis Jourdan
D, Alfred Hitchcock*

Hitchcock in uncharacteristic vein with a
courtroom drama; Valli is on trial for
murdering her husband and falls in love
with defending advocate Peck. Laughton's
coarse judge gives sparkle to an otherwise
routine affair.
Guild   VHS, Beta   A

The Parallax View*
US 1974 101m v colour

*Warren Beatty, Paula Prentiss, William
Daniels, Hume Cronyn, Walter McGinn
D, Alan J Pakula*

Witnesses to the assassination of an
American senator are killed one by one and
crusading journalist Beatty is determined
to find out why. Intense and intricate
psychological thriller by the director of
Klute (qv).
CIC   VHS, Beta, V2000   B

Perfect Friday
GB 1970 92m v colour

*Stanley Baker, Ursula Andress, David
Warner, Patience Collier, T P McKenna
D, Peter Hall*

Baker decides to enliven his life by robbing
the bank where he is deputy manager;
Andress and Warner are his partners in
crime. Lightweight caper with an unlikely
director; takes time coming to the boil.
VCL   VHS, Beta   A

Point Blank*
US 1967 88m v colour

*Lee Marvin, Angie Dickinson, Keenan
Wynn, Carroll O'Connor, Lloyd Bochner
D, John Boorman*

British director Boorman gets underneath
the American skin with this chilling revenge
piece. A hoodlum (Lee Marvin) roams the
Los Angeles underworld tracking the gang
who left him for dead. Sometimes flashy,
mostly compelling.
MGM/UA   VHS, Beta   A

The Postman Always Rings Twice*

US 1981 122m ● colour

Jack Nicholson, Jessica Lange, John Colicos, Anjelica Huston
D, Bob Rafelson

Second Hollywood treatment of James M Cain's classic 1930s tale of lust, murder and revenge with the sex left in this time. Nicholson and Lange play the lovers; powerful period detail.
Guild VHS, Beta, V2000 A

The Private Life of Sherlock Holmes*

GB 1970 121m ▢ colour

Robert Stephens, Colin Blakely, Clive Revill, Genevieve Page, Christopher Lee
D, Billy Wilder

Holmes is confronted by a Russian ballerina and unmasks a German spy in Scotland. The two adventures are not from the pen of Conan Doyle, but are executed with an affectionate sense of pastiche.
Warner VHS, Beta A

The Public Enemy**

US 1931 82m v bw

James Cagney, Edward Woods, Jean Harlow, Joan Blondell, Beryl Mercer
D, William Wellman

The classic tale of the rise and fall of a gangster, born on the wrong side of the tracks, building an empire in bootleg liquor and getting his come-uppance at the hands of a rival mob. The film that made a star of James Cagney.
Warner VHS, Beta A

Puppet on a Chain

GB 1970 98m v colour

Sven-Bertil Taube, Barbara Parkins, Patrick Allen, Alexander Knox
D, Geoffrey Reeve and Don Sharp

American Interpol man smokes out a gang of heroin smugglers in Amsterdam. Nasty and brutish piece, written specially for the screen by Alistair MacLean; the speedboat chase through the canals comes as wholesome relief.
Guild VHS, Beta, V2000 A

The Riddle of the Sands

GB 1978 102m ▢ colour

Michael York, Simon MacCorkindale, Jenny Agutter, Alan Badel
D, Tony Maylam

York as the English yachtsman holidaying in the North Sea who stumbles on a German plan to invade Britain. Muted version of Erskine Childers' classic spy story that looks the part but fails to set the pulses racing.
Rank VHS, Beta, Laser A

Sabotage*

GB   1936   73m   v   bw

*Oscar Homolka, Sylvia Sidney, John
Loder, Desmond Tester
D, Alfred Hitchcock*

Loose adaptation of Joseph Conrad's *The
Secret Agent*, with Homolka as the anar-
chist planning to destroy London. Contains
two vintage Hitchcock sequences: the boy
with the time-bomb and Sylvia Sidney with
a carving knife.

Rank   VHS, Beta   A

The St Valentine's Day Massacre*

US   1967   97m   ●   colour

*Jason Robards Jnr, George Segal, Ralph
Meeker, Jean Hale, Clint Ritchie
D, Roger Corman*

Re-creation of the Chicago gang wars of
the 1920s, with Al Capone (Robards)
versus Bugsy Moran (Meeker). Corman
treats it in cool style with crisp characterisation and care for period.

CBS/Fox   VHS, Beta   A

Sapphire

GB   1959   92m   v   colour

*Nigel Patrick, Michael Craig, Yvonne
Mitchell, Paul Massie, Bernard Miles
D, Basil Dearden*

Detective Nigel Patrick investigates the
murder of a coloured girl in London. Briskly
directed whodunit; unusual and bold for its
time in raising the issue of race; graced by a
crop of solid performances.

Rank   VHS, Beta   A

Serpico*

US   1973   126m   ●   colour

*Al Pacino, John Randolph, Jack Kehoe,
Biff McGuire
D, Sidney Lumet*

Pacino, as the New York cop with a
conscience, refuses to go along with the
corruption in the force and finds himself
isolated. Vivid re-enactment of a true case,
shot on the streets where it actually
happened.

CIC   VHS, Beta   A

The Seven Ups

US   1973   99m   ●   colour

*Roy Scheider, Victor Arnold, Jerry Leon,
Tony Lo Bianco, Richard Lynch
D, Philip D'Antoni*

Archetypal 1970s dirty cop movie, with
Scheider leading his bunch of police
roughnecks against a couple of thugs who
are kidnapping rival hoodlums and holding
them to ransom. Lots of violent action and
the inevitable car chase.

CBS/Fox   VHS, Beta, Laser   A

Shaft*
US 1971 96m ● colour

Richard Roundtree, Moses Gunn,
Charles Cioffi, Christopher St John
D, Gordon Parks

A private eye takes on the Mafia who are
muscling in on the Harlem rackets. Slick
thriller with plenty of pace; made by a black
director with a mainly black cast. Also
available: *Shaft in Africa, Shaft's Big Score*
MGM/UA VHS, Beta A

Shamus
US 1972 96m ● colour

Burt Reynolds, Dyan Cannon, John Ryan,
Joe Santos, Giorgio Tozzi
D, Buzz Kulik

Private dick Reynolds wanders through a
convoluted plot about murder and stolen
jewels. Jokey nod towards *The Big Sleep*
(qv), laced with 1970s sex and violence; but
Reynolds and Cannon are no Bogart and
Bacall.
RCA/Columbia VHS, Beta A

Spellbound**
US 1945 111m v bw

Ingrid Bergman, Gregory Peck, Leo G
Carroll, Michael Chekhov, Rhonda
Fleming
D, Alfred Hitchcock

A doctor in a mental asylum (Bergman)
unravels the guilty secret of her new boss
(Peck). Teasing Hitchcock suspense with a
strong dash of Freud; dream sequences by
Salvador Dali; lush, Oscar-winning score.
Guild VHS, Beta A

The Spiral Staircase**
US 1945 83m v bw

Dorothy McGuire, George Brent, Kent
Smith, Ethel Barrymore, Rhys Williams
D, Robert Siodmak

A spooky mansion, a thunderstorm and
deaf mute Dorothy McGuire as the next
target for a psychopathic killer at large in a
New England town. Just the ingredients to
make the flesh creep; director Siodmak
mixes them brilliantly.
Guild VHS, Beta A

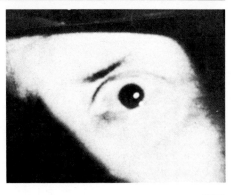

COBURN
RRYMORE
RDAN
VALLI

Directed by
ALFRED HITCHCOCK
Produced by
DAVID O. SELZNICK

The Paradine Case

Public Enemy

The St Valentine's Day Massacre

The Spy Who Loved Me
GB 1977 125m v colour
Roger Moore, Barbara Bach, Curt Jurgens, Richard Kiel, Caroline Munro
D, Lewis Gilbert

Megalomaniac Jurgens trying to destroy the world and build a new one under the sea; but 007 and the glamorous Anya are equal to the challenge. Standard James Bond fare enlivened by Richard Kiel as the seven-foot heavy, Jaws.
Warner VHS, Beta A

Still of the Night
US 1982 88m v colour
Roy Scheider, Meryl Streep, Jessica Tandy, Joe Grifasi, Sara Botsford
D, Robert Benton

Psychiatrist Roy Scheider has a murder on his hands and the dead man's mistress (Streep) knows more than she is letting on. Lightweight sub-Freudian thriller, littered with dreams, mother-complexes and childhood traumas.
Warner VHS, Beta A

The Sting**
US 1973 124m v colour
Paul Newman, Robert Redford, Robert Shaw, Charles Durning
D, George Roy Hill

Newman and Redford, as Chicago conmen of the 1930s, plot ingenious revenge on the mobster who killed their buddy. Shrewdly packaged entertainment, built on star appeal and period colour. Also available: *The Sting II*
CIC VHS, Beta, Laser A

The Stone Killer
US 1973 94m ● colour
Charles Bronson, Martin Balsam, Ralph Waite, David Sheiner, Norman Fell
D, Michael Winner

Dirty cop Charles Bronson tangles with the Mafia in Los Angeles. Violent, fast-moving and indulgently directed tour of movie gangland, taking in drug pushers, black militants, Vietnam veterans and the indispensable car chase.
RCA/Columbia VHS, Beta A

Strangers on a Train***
US 1951 101m v bw
Farley Granger, Robert Walker, Ruth Roman, Leo G Carroll
D, Alfred Hitchcock

"You do my murder and I will do yours", says mad Bruno (Walker) to tennis ace Granger; and Bruno is not joking. A giant among Hitchcock thrillers which unfolds with relentless logic and grips to the very last frame.
Warner VHS, Beta A

Sweeney*
GB 1976 94m ● colour

John Thaw, Dennis Waterman, Barry Foster, Ian Bannen, Diane Keen
D, David Wickes

Flying Squad detectives Regan and Carter uncover an international oil conspiracy. Television spin-offs do not usually work in the cinema but this one has much of the caustic flavour of the original, helped by the central performances.
Thorn EMI   VHS, Beta   A

Sweeney II
GB 1978 103m ● colour

John Thaw, Dennis Waterman, Barry Stanton, John Flanagan, David Casey
D, Tom Clegg

Another case for Regan and Carter, to track down the perpetrators of a series of armed bank robbers. Coarser treatment than before, particularly in the use of language, with the emphasis on police brutality and corruption.
Thorn EMI   VHS, Beta   A

The Taking of Pelham 123*
US 1974 105m v colour

Walter Matthau, Robert Shaw, Martin Balsam, Hecktor Elizondo
D, Joseph Sargent

Drama on the New York subway as four gunmen hi-jack a train and demand a $1m ransom. Efficient nail-biting job by director Sargent, with Shaw and Balsam leading the heavies and Matthau as the cop.
Warner   VHS, Beta   A

Telefon
US 1977 99m v colour

Charles Bronson, Lee Remick, Donald Pleasence, Alan Badel, Patrick Magee
D, Don Siegel

A KGB man (Bronson) hopes to help East–West relations by eliminating Pleasence and a bunch of fanatics who are sabotaging military targets across the United States. Director Siegel can do better than this.
MGM/UA   VHS, Beta   A

10 Rillington Place
GB 1970 108m ● colour

Richard Attenborough, John Hurt, Judy Geeson, Pat Heywood, Isobel Black
D, Richard Fleischer

Creditably unsensationalised account of the life and crimes of the mass murderer, John Reginald Christie. Attenborough, with bald pate and heavy breathing, is upstaged by John Hurt as the pathetic Timothy Evans.
RCA/Columbia   VHS, Beta   A

The Third Man ***

GB 1949 99m v bw

*Joseph Cotten, Orson Welles, Alida Valli,
Trevor Howard, Wilfrid Hyde White*
D, *Carol Reed*

Alexander Korda wanted Graham Greene and the director Carol Reed to make another film to follow their success with *The Fallen Idol* (qv); he proposed a story set in Vienna during the four-power occupation; Greene remembered an idea he had written years before on an envelope about a man recently buried who apparently comes back to life. So *The Third Man* was born. It opens with Holly Martins, an American writer of hack Westerns, arriving in Vienna to meet his friend, Harry Lime. He is told that Harry has been killed in an accident; but all is not what it seems. The casting was impeccable: Orson Welles as the genial, evil Lime (writing in for himself the film's most memorable piece of dialogue, about the Swiss and cuckoo clocks); Cotten as Martins, the innocent adrift in a strange city; Valli as Harry's girl; and Howard as the laconic British military policeman. The baroque and war-scarred city of Vienna was brilliantly used by Reed and his cameraman, Robert Krasker. Reed was lucky to discover the zither player, Anton Karas, whose haunting tunes enrich the atmosphere of the film far more than any conventional score. The ending, insisted upon by Reed despite Greene's opposition, remains one of the finest in the cinema.

Thorn EMI VHS, Beta A

The Thirty-Nine Steps***
GB 1935 78m v bw

Robert Donat, Madeleine Carroll, Godfrey Tearle, Lucie Mannheim
D, Alfred Hitchcock

Marvellously witty and inventive double chase thriller with Donat, handcuffed to Madeleine Carroll, on the run from both the police and enemy agents. Not much of John Buchan but the Hitchcock touch more than compensates.
Rank VHS, Beta, Laser, CED A

The Thirty-Nine Steps
GB 1978 102m v colour

Robert Powell, Karen Dotrice, John Mills, Eric Porter, David Warner
D, Don Sharp

Another crack at the Buchan novel, with less flair than the Hitchcock version but set in the correct period (just before the First World War) and closer to the original story. Robert Powell as Hannay and a climax on the face of Big Ben.
Rank VHS, Beta A

The Thomas Crown Affair*
US 1968 102m v colour

Steve McQueen, Faye Dunaway, Paul Burke, Jack Weston, Yaphet Kotto
D, Norman Jewison

McQueen, as a bored young tycoon, masterminds a bank robbery and falls in love with the insurance company's pretty investigator. A glossy confection in which director Jewison puts style a long way before substance.
Warner VHS, Beta A

Thunderball*
GB 1965 128m v colour

Sean Connery, Adolfo Celi, Claudine Auger, Luciana Paluzzi
D, Terence Young

James Bond on the trail of two hi-jacked atom bombs engages in a desperate undersea battle with the villains of SPECTRE. One of the better examples of the 007 series, made while the formula was still fresh.
Warner VHS, Beta A

An Unsuitable Job for a Woman*
GB 1981 94m v colour

Billie Whitelaw, Pippa Guard, Paul Freeman, Dominic Guard
D, Christopher Petit

Pippa Guard as P D James' young private detective, Cordelia Gray, investigating the mysterious death of a student. Stylish treatment of a complex psychological whodunit, with excellent performances.
Palace VHS, Beta A

Villain*
GB 1971 91m ● colour

Richard Burton, Ian MacShane, Nigel Davenport, Joss Ackland
D, Michael Tuchner

Tightly plotted, crisply written (by Ian La Frenais and Dick Clement) and efficiently staged story of a homosexual East End gang boss getting his deserts during a payroll robbery. Definitely not for the squeamish.
Thorn EMI VHS, Beta A

The Wrong Man**
US 1957 101m v bw

*Henry Fonda, Vera Miles, Anthony
Quayle, Harold J Stone
D, Alfred Hitchcock*

Rare excursion by Hitchcock into semi-documentary, charting the nightmare experiences of a New York double-bass player (Fonda), arrested for a hold-up he did not commit. Based on a true story, with authentic locations.
Warner VHS, Beta A

While the City Sleeps*
US 1956 97m ● bw

*Dana Andrews, George Sanders, Vincent
Price, Rhonda Fleming, Thomas Mitchell
D, Fritz Lang*

Contest between three newspapermen – with promotion as the ''prize'' – to track down a sex murderer who is terrorising New York. A minor work in the Fritz Lang canon, though one of his favourites. With *Stromboli* (qv).
Kingston VHS, Beta A

White Heat***
US 1949 114m ● bw

*James Cagney, Edmond O'Brien,
Margaret Wycherly, Virginia Mayo
D, Raoul Walsh*

Cagney gives his most compelling gangster performance as the psychopathic, mother-fixated hoodlum who meets his spectacular end at a chemical plant. A non-stop explosion of raw energy, given full rein by director Walsh.
Warner VHS, Beta A

You Only Live Twice
GB 1967 116m v colour

*Sean Connery, Tetsuro Tamba, Akiko
Wakabayashi, Charles Gray, Donald
Pleasence
D, Lewis Gilbert*

James Bond swims ashore in Japan, on the trail of a missing American space capsule. The technology is starting to take over the series, though Pleasence is a suitably sneering villain.
Warner VHS, Beta A

Young and Innocent**
GB 1937 77m v bw

*Nova Pilbeam, Derrick de Marney, Mary
Clare, Edward Rigby, Basil Radford
D, Alfred Hitchcock*

A young man wrongly suspected of murder is pursued by the police, while he tries to find the real culprit. Engagingly unpretentious treatment of a favourite Hitchcock theme, the double chase, which still comes up fresh.
Rank VHS, Beta A

PAUL
ROBERT   ORD

and the

CASSIDY
UNDANCE KID

The Alamo*
US 1960 161m v colour

*John Wayne, Richard Widmark, Richard
Boone, Laurence Harvey, Frankie Avalon*
D, John Wayne

Wayne as the all-American hero, Davy
Crockett, perishing gloriously at the hands
of the Mexicans during the struggle for
Texan independence. Big-budget spectac-
ular with impressive battle scenes.
Warner VHS, Beta A

Alvarez Kelly
US 1966 v colour

*William Holden, Richard Widmark,
Janice Rule, Patrick O'Neal*
D, Edward Dymtryk

Irish-Mexican rancher (Holden) forced to
turn cattle rustler by Confederate guerilla
(Widmark) during the American Civil War.
Enjoyably meandering piece, with inter-
ludes for comedy and romance; builds to a
tense climax.
RCA/Columbia VHS, Beta A

Bandolero!
US 1968 102m ● colour

*James Stewart, Dean Martin, Raquel
Welch, George Kennedy, Will Geer*
D, Andrew V McLaglen

Sheriff George Kennedy in dogged pursuit
of outlaw brothers Stewart and Martin and
the shapely widow (Welch) they take
hostage on the way. Over-written but
efficiently handled with a pleasing eye for
offbeat detail.
CBS/Fox VHS, Beta, Laser A

Best of the Badmen
US 1951 81m □ colour

*Robert Ryan, Claire Trevor, Jack Buetel,
Robert Preston, Walter Brennan*
D, William D Russell

Ryan as a US cavalry officer arrested on a
trumped-up murder charge; he escapes
and organises an army of outlaws to seek
revenge. Standard Western fare with an
above-average cast. With: *Sealed Cargo*,
1951 adventure.
Kingston VHS, Beta A

The Big Sky*
US 1952 117m v bw

*Kirk Douglas, Arthur Hunnicutt, Elizabeth
Threatt, Dewey Martin*
D, Howard Hawks

The relationships between two Kentucky
mountain men (Douglas and Martin) and
an Indian girl during a pioneer trip up the
Missouri in the 1830s. A lesser work from
a great director. With: *Code of the West*,
1948 Western.
Kingston VHS, Beta A

Buffalo Bill and the Indians*
US 1976 118m ● colour

*Paul Newman, Burt Lancaster, Joel Grey,
Kevin McCarthy, Geraldine Chaplin*
D, Robert Altman

Subtitled *Sitting Bull's History Lesson* and
showing how the Western legend has been
manipulated to whitewash the white man.
Paul Newman, as Buffalo Bill Cody, takes
his Wild West Show across America in the
1880s.
Thorn EMI VHS, Beta A

Butch and Sundance: the Early Days*

US 1979 107m v colour

Tom Berenger, William Katt, Jeff Corey, John Schuck, Michael C Gwynne
D, Richard Lester

"Prequel" to *Butch Cassidy and the Sundance Kid* (see below) which eschews big star performances and replaces the easy-going charm of the original with a less romantic view of the protagonists.
CBS/Fox   VHS, Beta, V2000   A

Butch Cassidy (TOM BERENGER) on t engine of the mail train, during the c

Butch Cassidy and the Sundance Kid**

US 1969 106m v colour

Robert Redford, Paul Newman, Katharine Ross, Strother Martin, Henry Jones
D, George Roy Hill

Newman and Redford as turn of the century outlaws on the run in a cheerfully rose-tinted version of the Old West which spawned the Oscar-winning song, "Raindrops Keep Fallin' on My Head".
CBS/Fox   VHS, Beta, V2000, Laser   A

Cahill

US 1973 97m v colour

John Wayne, George Kennedy, Gary Grimes, Neville Brand, Clay O'Brien
D, Andrew V McLaglen

Wayne as an old marshal trying to bridge a generation gap when his sons kick over the traces and take part in a robbery. The lads are a little too cute and Kennedy overdoes the heavy, leaving Big John in quiet control.
Warner   VHS, Beta   A

Chisum

US 1970 110m □ colour

John Wayne, Forrest Tucker, Christopher George, Ben Johnson, Glenn Corbett
D, Andrew V McLaglen

Cattle baron John Chisum (Wayne) fights to save his ranch from the evil Forrest Tucker, with the help of friends Pat Garrett and Billy the Kid. Competently handled, stylishly photographed reworking of old themes.
Warner   VHS, Beta   A

Comes a Horseman*

US 1978 119m v colour

Jane Fonda, Jason Robards Jnr, James Caan, George Grizzard
D, Alan J Pakula

Intriguing modern Western, set towards the end of the Second World War, with Caan and Fonda as Montana ranchers trying to save their land from rival Robards and an oilman. An elegaic study of people whom time is passing by.
Warner   VHS, Beta   A

Buffalo Bill and the Indians

Comes a Horseman

For a Few Dollars More

The Culpepper Cattle Company*

US   1972   92m   ●   colour

Gary Grimes, Billy 'Green' Bush, Luke Askew, Bo Hopkins, Geoffrey Lewis
D, Dick Richards

The harsh initiation of a 16-year-old would-be cowboy (Grimes) into the violent ways of the world during a cattle drive to Colorado. Relentlessly sombre treatment of the Western myth, minus the glamour.
CBS/Fox   VHS, Beta, V2000, Laser   A

Drum Beat*

US   1954   105m   □   colour

Alan Ladd, Audrey Dalton, Marisa Pavan, Robert Keith, Charles Bronson
D, Delmer Daves

Alan Ladd seeks peace with the Modoc Indians during the 1860s; but renegade Bronson refuses to yield. Writer-director Daves helps to give it a ring of authenticity. With: *To Beat the Band*, 1935 comedy-musical.
Kingston   VHS, Beta   A

Duel in the Sun**

US   1946   130m   v   colour

Jennifer Jones, Joseph Cotten, Gregory Peck, Lionel Barrymore, Lillian Gish
D, King Vidor

A lavish attempt at a Western *Gone with the Wind* (qv), soon christened "Lust in the Dust". Set on a Texas ranch, with Jennifer Jones as a half-breed who comes between two brothers; over-cooked and bitty but still compelling.
Guild   VHS, Beta, V2000   A

Eagle's Wing*

GB   1979   104m   v   colour

Martin Sheen, Sam Waterston, Harvey Keitel, Stephane Audran, John Castle
D, Anthony Harvey

White trapper (Sheen) and Comanche chief (Waterston) dispute possession of a prize white stallion. Quirky, symbolic piece made by a British director in Spain, with magnificent photography but uncertain about its real aim.
Rank   VHS, Beta   A

A Fistful of Dollars*

Italy/WG/Spain 1964 95m ● col

Clint Eastwood, Gian Maria Volonte, Marianne Koch, Pepe Calvo
D, Sergio Leone

Eastwood as the cheroot-smoking, unnamed stranger who cleans up a Mexican border town. He launches himself towards international stardom in the first of the moody, violent genre which has become known as the "spaghetti" Western.
Warner   VHS, Beta   A

For a Few Dollars More

Ita/Spa/WG 1965 133m ● colour

Clint Eastwood, Lee Van Cleef, Gian Maria Volonté, Klaus Kinski
D, Sergio Leone

Long drawn out sequel to *A Fistful of Dollars* (qv) with Eastwood re-creating his mysterious stranger and joining forces with a bounty hunter (Van Cleef) to run down a deranged killer. Unappetising fare but efficiently served.
Warner   VHS, Beta   A

The Good, the Bad and the Ugly

Italy   1966   161m ● colour

Clint Eastwood, Eli Wallach, Lee Van Cleef
D, Sergio Leone

Confidence tricksters Eastwood and Wallach and murderer Van Cleef look for $200,000 hidden in a cemetery during the American Civil War. Another helping of Leone-style spaghetti.
Warner   VHS, Beta   A

A Gunfight

US   1970   97m   v   colour

Kirk Douglas, Johnny Cash, Karen Black, Raf Vallone, Jane Alexander
D, Lamont Johnson

Douglas and Cash as two old gunfighters down on their luck who decide to fight a duel for money. The result is anti-climax, except that the film makers provide an alternative ending to make you sit up and think.
Rank   VHS, Beta   A

Hang 'em High

US   1967   114m ● colour

Clint Eastwood, Inger Stevens, Ed Begley, Pat Hingle, James MacArthur
D, Ted Post

The spaghetti Western goes to Hollywood. Eastwood stays in the saddle, as a cowboy who escapes a lynching and wreaks vengeance on the nine men who set him up. Has all the brutality of the genre but not its redeeming style.
Warner   VHS, Beta   A

Hannie Caulder

GB   1971   85m ● colour

Raquel Welch, Robert Culp, Ernest Borgnine, Strother Martin, Jack Elam
D, Burt Kennedy

Raquel Welch doing her bit for the feminist cause by getting after the three unspeakables (Borgnine, Martin, Elam) who raped her after killing her husband. Best taken with a fistful of salt.
Video Movies   VHS, Beta   A

Heaven's Gate*
US   1980   148m   ●   colour

Kris Kristofferson, Christopher Walken,
John Hurt, Sam Waterston
D, Michael Cimino

Cimino's *folie de grandeur*, a $36m recrea-
tion of the Johnson County War between
cattlemen and immigrant farmers in the
1890s. Blasted by the critics, withdrawn
and re-edited but worth a look all the same.
Warner   VHS, Beta   A

High Plains Drifter
US   1972   101m   ●   colour

Clint Eastwood, Verna Bloom, Marianna
Hill, Mitch Ryan, Jack Ging
D, Clint Eastwood

Eastwood turns director and picks up a hint
or two from his mentor, Sergio Leone,
about moody atmospherics and eruptions
of violence. Eastwood the actor is back in
his mysterious stranger role, terrifying a
town.
CIC   VHS, Beta   A

Jeremiah Johnson*
US   1972   104m   v   colour

Robert Redford, Will Geer, Alyn Ann
McLerie, Stephan Gierasch
D, Sydney Pollack

Redford as an ex-soldier who gets the call
of the wild and becomes a mountain
trapper and is forced into a vendetta
against the killers of his Indian wife.
Episodic, quasi-allegorical; dazzling snow-
capes.
Warner   VHS, Beta   A

Joe Kidd
US   1972   87m   v   colour

Clint Eastwood, Robert Duvall, John
Saxon, Don Stroud
D, John Sturges

Strong, taciturn Eastwood getting involved
in New Mexican land wars, tangling with
land baron Duvall and Mexican bandit
Saxon. Essentially a vehicle for its star, set
against the imposing backcloth of the High
Sierras.
CIC   VHS, Beta   B

Kid Blue*
US   1973   97m   v   colour

Dennis Hopper, Warren Oates, Peter
Boyle, Ben Johnson, Lee Purcell, Janice
Rule
D, James Frawley

Hopper as a young Texan outlaw who tries
to settle down and get a job but finds that
respectability does not work. Amusingly
presented study of a society unprepared to
accept a non-conformer, with neat touches.
CBS/Fox   VHS, Beta   A

Little Big Man*
US 1970 134m v colour

*Dustin Hoffman, Martin Balsam, Faye
Dunaway, Chief Dan George
D, Arthur Penn*

A tragi-comic look at the Indian wars of the
nineteenth century as recalled by a 121-
year-old survivor (Hoffman). Picturesque
attempt to re-write American history by
showing that the white man could be
wrong.
CBS/Fox VHS, Beta A

The Long Riders
US 1980 100m ● colour

*Stacy and James Keach; David, Keith
and Robert Carradine; Dennis and Randy
Quaid
D, Walter Hill*

The exploits of those favourite outlaw
brothers – Frank and Jessie James, the
Youngers and the Millers – with real
brothers playing them. A clever idea, which
is confused by director Hill's handling.
Warner VHS, Beta A

The Magnificent Seven*
US 1960 138m v colour

*Yul Brynner, Steve McQueen, James
Coburn, Charles Bronson, Robert Vaughn
D, John Sturges*

Pinching their plot from Kurosawa's
Japanese classic, *The Seven Samurai*,
Brynner and friends ride into Mexico to
protect a village against bandits. Plenty of
uncomplicated action and a stirring score
by Elmer Bernstein.
Warner VHS, Beta A

Major Dundee*
US 1964 ● colour

*Charlton Heston, Richard Harris, Jim
Hutton, James Coburn, Michael
Anderson Jnr
D, Sam Peckinpah*

Tough Federal officer Charlton Heston
leading a bunch of Confederate prisoners
of war (under Richard Harris) against the
Apaches. Allegedly mangled after leaving
Peckinpah's hands but still absorbing.
RCA/Columbia VHS, Beta A

The Missouri Breaks
US 1976 126m v colour

*Marlon Brando, Jack Nicholson, Randy
Quaid, Kathleen Lloyd, Frederic Forrest
D, Arthur Penn*

Dense, eccentric tale of cattle rustling in
Montana in the 1880s which is dominated
by Brando's extravagant performance as
the man hired to sort out a gang boss
(Nicholson). Director Penn seems bemused
by it all.
Warner VHS, Beta A

The Outlaw*
US 1943 111m □ bw

*Jack Buetel, Jane Russell, Thomas
Mitchell, Walter Huston
D, Howard Hughes*

The film that launched Jane Russell and
her ample bosom; a moody version of Billy
the Kid (Buetel), with Mitchell as Pat
Garrett and Huston as Doc Holliday. With:
Law of the Underworld, 1938 thriller.
Kingston VHS, Beta A

The Outlaw Josey Wales*
US 1976 135 v colour

Clint Eastwood, Chief Dan George,
Sondra Locke, John Vernon
D, Clint Eastwood

Missouri farmer Eastwood joins a Southern guerilla band during the American Civil War to avenge the murder of his wife and child and acquires a new family on the way. Eastood as director handles with a sure touch.
Warner VHS, Beta A

Pat Garrett and Billy the Kid
US 1973 102m ● colour

James Coburn, Kris Kristofferson, Bob
Dylan, Richard Jaeckel
D, Sam Peckinpah

Intense, downbeat treatment of two men on their way to a tragic destiny, the sheriff (Coburn) and the one time friend (Kristofferson) he is forced to kill. Patches of incoherence, perhaps reflecting production troubles.
MGM/UA VHS, Beta, V2000 A

Once Upon a Time in the West*
Italy/US 1968 159m ● colour

Henry Fonda, Claudia Cardinale, Jason
Robards, Charles Bronson, Frank Wolff
D, Sergio Leone

Shady lady Cardinale finds husband and family murdered; outlaws Robards and Bronson rally round; Fonda, for once in his career, is the baddie. Long, but absorbing with lashings of violence and sumptuous camerawork.
CIC VHS, Beta, V2000 B

Rio Bravo***
US 1959 137m □ colour

John Wayne, Dean Martin, Ricky Nelson,
Angie Dickinson, Walter Brennan
D, Howard Hawks

Sheriff Wayne, reformed drunk Martin, old-timer Brennan and greenhorn Nelson have to defend a town against outlaws; but director Hawks is less interested in plot than the interplay of character. Among the best Westerns ever.
Warner VHS, Beta A

Run of the Arrow**
US 1957 83m ● colour

Rod Steiger, Sarita Montiel, Charles
Bronson, Tim McCoy, Ralph Meeker
D, Samuel Fuller

Steiger as a deserter from the Confederate army who is captured by the Sioux Indians and marries one of them but finally rejects their way of life; a trenchant insight into cultural differences. With: *Cry Danger* (qv).
Kingston VHS, Beta A

Shalako
GB 1968 108m v colour

Sean Connery, Brigitte Bardot, Jack Hawkins, Stephen Boyd, Peter Van Eyck
D, Edward Dmytryk

Connery takes a break from Bond to play a cowboy trying to save a group of European big game hunters from the Apaches. Impressive locations with Spain standing in for New Mexico, but fails to raise the pulse-rate.
Thorn EMI   VHS, Beta   A

Shane***
US 1953 113m □ colour

Alan Ladd, Jean Arthur, Van Heflin, Jack Palance, Brandon de Wilde
D, George Stevens

Alan Ladd as the mysterious stranger, riding into town and ridding it of the treacherous Jack Palance. Slow-moving but perfectly controlled film, with magnificent colour photography and a host of memorable set pieces.
CIC   VHS, Beta   A

Stagecoach***
US 1939 92m v bw

Claire Trevor, John Wayne, Thomas Mitchell, John Carradine, Andy Devine
D, John Ford

Stagecoach is based on that age-old dramatic ploy of putting a bunch of disparate characters into a common predicament and seeing how they make out. The passengers on the stage know that it may be attacked by Indians en route from Tucson to New Mexico. On board are Dallas (Trevor), the prostitute hounded out of town by the ladies of the Law and Order League; the upright Mrs Mallory (Louise Platt), a cavalry officer's wife heavily pregnant; the drunken Doc Boone (Mitchell); and Gatewood (Berton Churchill), absconding with bank funds. On the way they pick up the Ringo Kid (Wayne),

bent on avenging the gang who have framed him for murder. The attack by Geronimo's Apaches is brilliantly staged against what was to become John Ford's most famous location, Monument Valley, Utah, and it is the common danger against which all can unite. This is not an innovative film; Ford is content to work within established conventions and to re-work familiar themes. But he does so with deceptive skill and his characters are all beautifully delineated within the first few minutes. The narrative control is firm throughout; there is room for sentiment and for the typically robust Fordian sense of humour. The style is simple, unaffected and absolutely right. For many *Stagecoach* is still *the* Western, setting a standard for the rest. With: *Deadline at Dawn* (qv).
Kingston   VHS, Beta   A

275

There was a Crooked Man*
US 1970 126m v colour

*Kirk Douglas, Henry Fonda, Hume
Cronyn, Warren Oates, Burgess Meredith
D, Joseph L Mankiewicz*

Witty, intelligent, ironic tale of a bespectacled convict (Douglas) who manipulates his fellow prisoners and tries to hoodwink the sheriff (Fonda) who put him inside; splendid character work from the supporting cast.
Warner   VHS, Beta   A

Tom Horn
US 1980 93m v colour

*Steve McQueen, Linda Evans, Richard
Farnsworth, Bill 'Green' Bush
D, William Wiard*

The more or less true story of the ex-cavalry scout and bounty hunter who is given the job of flushing out cattle rustlers, carries his brief too far and is framed for murder. Subdued performance by McQueen in his penultimate film.
Warner   VHS, Beta   A

The Train Robbers
US 1973 88m □ colour

*John Wayne, Ann-Margret, Rod Taylor,
Ben Johnson, Bobby Vinton
D, Burt Kennedy*

Widow Ann-Margret asks old gunman Wayne and his friends to recover gold stolen by her train-robber husband. Director Kennedy tries to play it for laughs but no one is really in the mood; neat twist at the end, though.
Warner   VHS, Beta   A

True Grit*
US 1969 128m v colour

*John Wayne, Kim Darby, Glen Campbell,
Dennis Hopper, Jeremy Slate
D, Henry Hathaway*

John Wayne, with black eye-patch and a sizeable paunch, at last won an Oscar as Marshal Rooster Cogburn, helping a young woman (Darby) avenge the murder of her father. Old-fashioned showcase for a great Western star.
CIC   VHS, Beta   A

Wagonmaster***
US 1950 82m □ bw

*Ben Johnson, Joanne Dru, Harry Carey
Jnr, Ward Bond, Charles Kemper
D, John Ford*

Ford's hymn to the pioneer spirit, a lovingly assembled tribute to the values of the Old West as a Mormon wagon train fends off Indians and outlaws on its way to Utah in 1879. With: *Double Dynamite*, 1951 comedy.
Kingston   VHS, Beta   A

Price guide
VHS and Beta

A   £30 or more
B   £20–£30
C   Below £20
R   Rental only

Note: There are no fixed prices for video tapes and there may be variations from one dealer to another. V2000 tapes tend to be in the B price range, while discs are much cheaper with all titles at approximately £12.

Pat Garrett and Billy the Kid

Shane

True Grit

Hannie Caulder

Joe Kidd

The Outlaw

The Wild Bunch***
US 1969 138m ● colour

William Holden, Ernest Borgnine, Robert Ryan, Edmond O'Brien, Warren Oates
D, *Sam Peckinpah*

The taming of the West is one of the most potent of American myths, a celebration of enterprise and dynamism and the frontier spirit; and nowhere has the myth been more assiduously perpetuated than in Hollywood. However, *The Wild Bunch* shows a very different sort of West from this romantic vision. Peckinpah's people are vicious outlaws, who live only by violence. He deliberately sets his film as late as 1914, when motor cars are taking over from horses, to underline the fact that the Wild Bunch are men out of their time. As their leader (Holden) says to his sidekick: "We're getting old. We've got to think beyond our guns. I'd like to make one good score and back off." To which Borgnine replies: "Back off to what?" The film follows the Wild Bunch along their bloody trail from Texas and into Mexico, deeper and deeper into self-destruction. Almost for the first time in a Western, the violence is explicit. Whereas in the typical Western people die cleanly, without blood or pain, Peckinpah supplies both, slowing up the action to make the point more vividly. "I wanted to show," he said, "what the hell it feels like to get shot." Here is not the gratuitous slaughter of some of Peckinpah's other films but an artist making a striking statement. To some *The Wild Bunch* is an allegory on Vietnam; as a Western in its own terms it questions more cogently than any previous film in the genre the validity of a society based on gun rule; no longer can the world be made safe by the ritualistic shoot-out at high noon.
Warner VHS, Beta A

279

Best Family Films

The following films should appeal to people of all ages. Those films given a star rating are included, plus the best of the rest.

Action/adventure
The Adventures of Robin Hood***
The Black Pirate**
Captain Blood*
The Four Feathers*
The Thief of Baghdad (1940)***
The Three Musketeers*
The Wooden Horse*

Children
The Adventures of Tom Sawyer**
Chitty Chitty Bang Bang*
The Incredible Journey*
Jason and the Argonauts*
The Love Bug***
Mary Poppins***
One of Our Dinosaurs is Missing*
The Railway Children***
The Swiss Family Robinson*
That Darn Cat*
The Wizard of Oz***

Comedy
Cat Ballou**
Charlie Chaplin I**, II**, III**, IV**
City Lights***
The General***
Genevieve***
The Gold Rush***
The Great Race*
It's a Mad, Mad, Mad, Mad World*
The Lavender Hill Mob***
The Maggie*
Modern Times***
Monsieur Hulot's Holiday***
On the Beat/Trouble in Store
The Pink Panther*
Those Magnificent Men in their Flying
   Machines*
Whisky Galore***

Drama
Animal Farm
Chariots of Fire***
Oliver Twist***
Popeye
The Red Shoes***
Watership Down*

Musicals
Annie*
Bugsy Malone**
Dr Dolittle
The King and I*
My Fair Lady*
Oklahoma*
Oliver*
Seven Brides for Seven Brothers***
Singin' in the Rain***
The Sound of Music***

Science Fiction/Fantasy
Flash Gordon*
The Land That Time Forgot*
Lord of the Rings
Star Wars**
Superman*
Superman II*
Superman III*
20,000 Leagues Under the Sea*
Voyage to the Bottom of the Sea

Thrillers
The Face of Fu Manchu*

Westerns
Shane***
Stagecoach***

WARNER HOME VIDEO

ALEXANDER SALKIND PRESENTS MARLON BRANDO · GENE HACKMAN IN A RICHARD DONNER FILM
SUPERMAN

STARRING   ALSO STARRING
CHRISTOPHER REEVE · NED BEATTY · JACKIE COOPER · GLENN FORD · TREVOR HOWARD
MARGOT KIDDER · VALERIE PERRINE · MARIA SCHELL · TERENCE STAMP · PHYLLIS THAXTER · SUSANNAH YORK
STORY BY MARIO PUZO · SCREENPLAY BY MARIO PUZO, DAVID NEWMAN, LESLIE NEWMAN AND ROBERT BENTON
CREATIVE CONSULTANT TOM MANKIEWICZ · DIRECTOR OF PHOTOGRAPHY GEOFFREY UNSWORTH B.S.C.
PRODUCTION DESIGNER JOHN BARRY · MUSIC BY JOHN WILLIAMS · EXECUTIVE PRODUCER ILYA SALKIND
PRODUCED BY PIERRE SPENGLER · DIRECTED BY RICHARD DONNER · PANAVISION TECHNICOLOR
AN ALEXANDER AND ILYA SALKIND PRODUCTION

FROM WARNER BROS. A WARNER COMMUNICATIONS COMPANY

The 20 Classics

Action/Adventure
El Cid
Patton: Lust for Glory

Children
The Wizard of Oz

Comedy
City Lights
Kind Hearts and Coronets

Drama
Casablanca
Gone with the Wind
La Règle du Jeu
Wild Strawberries

Horror
Frankenstein (1931)
Psycho

Musicals
Singin' in the Rain
The Sound of Music

Sci-Fi/Fantasy
Metropolis
2001: A Space Odyssey

Thrillers
Bonnie and Clyde
The Maltese Falcon
The Third Man

Westerns
Stagecoach
The Wild Bunch

James Bond films

Reviews appear in the section on Thrillers
Diamonds Are Forever*
Doctor No*
For Your Eyes Only
From Russia with Love**
Goldfinger**
Live and Let Die
The Man with the Golden Gun
Moonraker
Never Say Never Again*
Octopussy
On Her Majesty's Secret Service
The Spy Who Loved Me
Thunderball*
You Only Live Twice

Pot luck

Here is a list of some outstanding and interesting films which were issued on video but for various reasons were later withdrawn. Because they are no longer in distributors' catalogues, we have not included them in the main body of the book; but it is possible that some copies may still be available through retail outlets. To give an idea of quality, we have used the same star system as in the rest of the book.

Action/adventure
One of Our Aircraft is Missing *

Comedy
Bringing Up Baby ***
Oh, Mr Porter ***

Drama
All About Eve ***
Citizen Kane ***
The Hustler **
Room At the Top *
The Trial **
Twelve O'Clock High *
Victoria the Great *

Horror
The Cabinet of Dr Caligari **
The Hunchback of Notre Dame (1939) **
King Kong (1933) ***

Musicals
Top Hat ***

Science fiction/fantasy
Invasion of the Body Snatchers (1956) **

Thrillers
Crossfire **
Laura ***

Westerns
Fort Apache **
High Noon **
Johnny Guitar *
The Lusty Men *
The Professionals *
The Searchers ***

An **MGM/CBS HOME VIDEO** Presentation

MGM

A STANLEY KUBRICK PRODUCTION

2001:
a space odyssey

Starring **KEIR DULLEA** · **GARY LOCKWOOD**
Director of Photography GEOFFREY UNSWORTH B.S.C.
Screenplay by STANLEY KUBRICK and ARTHUR C. CLARKE
Directed and Produced by STANLEY KUBRICK

Labels and distributors

Label	Distributor
Capstan	Capstan Video 266 Fulham Road London SW10 01-351-5059
CBS/Fox	CBS/Fox Video Perivale Industrial Park Greenford Middlesex UB6 7RU 01-997-2552
CIC	CIC Video UIP House Beadon Road Hammersmith London W6 01-741-9333
Embassy	Embassy Home Entertainment Holbein Place Sloane Square London SW1 01-730 3455
Entertainment	Entertainment Film Distributors Ltd National House 60-66 Wardour Street London W1 01-734 4678/9
Guild	Guild Home Video Ltd Guild House Oundle Road Peterborough PE2 9PZ (0733) 63122
Hokushin	Hokushin Audio Visual Ltd 2 Ambleside Avenue London SW16 01-769 0965
Intervision	Intervision Video Ltd Unit 1 McKay Trading Estate Kensal Road London W10 01-960 8211
Iver	Iver Film Services Ashley House 30 Ashley Road Altrincham Cheshire WA14 2DW 061-928 9011
Kingston	Kingston Video c/o Harris Films Ltd Glenbuck Road Surbiton Surrey 01-399 0022
Longman	Longman Video 21-27 Lamb's Conduit Street London WC1 01-242 2548
MEVC	Motion Epics Video Co Ltd 122/124 Regent Street London W1 01-434 3751
MGM/UA	MGM/UA Home Video Hammer House 113-117 Wardour Street London W1 01-439 9932
Palace	Palace, Virgin & Gold Distribution Ltd 69 Flempton Road London E10 01-539 5566
Precision	Precision Video Ltd ACC House 17 Gt Cumberland Place London W1 01-262 8040
Polygram	Polygram Video 1 Rockley Road London W14 01-743 3474

j.R.R. tolkien's

"the Lord of the Rings"

A SAUL ZAENTZ Production
A RALPH BAKSHI Film
J. R. R. TOLKIEN'S "THE LORD OF THE RINGS"
Music by LEONARD ROSENMAN
Screenplay by CHRIS CONKLING and PETER S. BEAGLE
Based on the novels *The Fellowship of the Ring*
and *The Two Towers* of J. R. R. TOLKIEN
Produced by SAUL ZAENTZ, Directed by RALPH BAKSHI

Labels and distributors

Rank	Rank Video Library PO Box 70 Great West Road Brentford Middlesex TW8 9HR 01-568 9222	Worldwide Entertainment	Worldwide Entertainment 38 Dover Street London W1 01-499 9416
RCA/Columbia	RCA/Columbia Pictures 1 Bedford Avenue London WC1 01-636 8311	Alpha Arena Brent Walker	CBS/Fox CIC Videospace
Thorn EMI	Thorn EMI Video Programmes Ltd Thorn EMI House Upper St Martins Lane London WC2 01-836 2444	Canon	Video Programme Distributors Ltd Building No. 1 GEC Estate East Lane Wembley Middlesex HA9 7QB
Videoform	Videoform Unit 4 Brunswick Industrial Park New Southgate London N11 01-368 1226	Disney Home Video Productions	01-904 0921 Rank Videospace
Videomedia	Videomedia 68/70 Wardour Street London W1 01-437 1563	Inter-Ocean Media	Video Programme Distributors Ltd Videoform
Videospace	Videospace Ltd 272 London Road Wallington Surrey SM6 7DJ 01-773 0921	Odyssey Spectrum 3M Video	CBS/Fox Polygram Video Palace, Virgin & Gold Distribution Ltd
Video Tape Center (VTC)	VTC plc No. 1 Newton Street London WC2 01-405 8484	Vampix VCL	Videomedia CBS/Fox
Warner	Warner Home Video PO Box 59 Alperton Lane Wembley Middlesex HA0 1FJ 01-998 8844	Virgin	Palace, Virgin & Gold Distribution Ltd

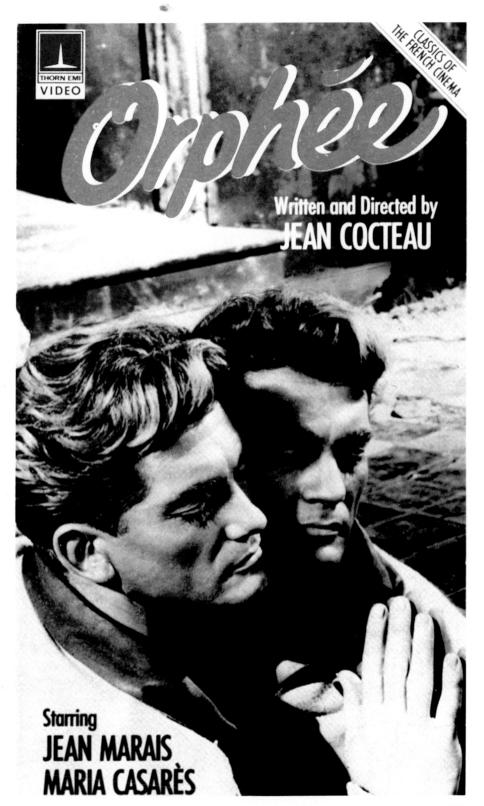

287

Index of films

Page No

WARNER HOME VIDEO

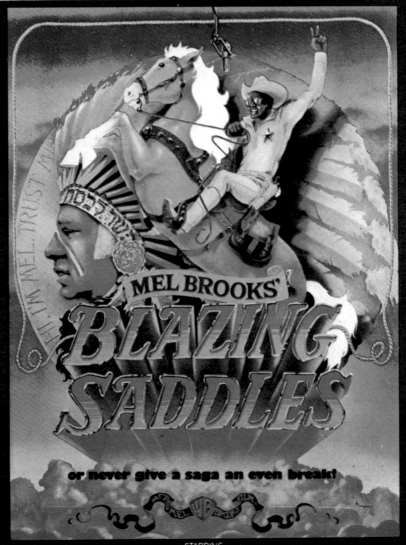

MEL BROOKS'

BLAZING SADDLES

or never give a saga an even break!

STARRING
CLEAVON LITTLE • GENE WILDER • SLIM PICKENS
DAVID HUDDLESTON • CLAUDE ENNIS STARRETT, JR.
ALSO STARRING
MEL BROOKS, HARVEY KORMAN and MADELINE KAHN
SCREENPLAY BY MEL BROOKS NORMAN STEINBERG ANDREW BERGMAN RICHARD PRYOR ALAN UGER
STORY BY ANDREW BERGMAN • PRODUCED BY MICHAEL HERTZBERG • DIRECTED BY MEL BROOKS

FROM WARNER BROS. A WARNER COMMUNICATIONS COMPANY

Throughout history he has filled the hearts of men with *terror,* and the hearts of women with *desire.*

DRACULA

1979 VERSION

THE MIRISCH CORPORATION PRESENTS FRANK LANGELLA WITH LAURENCE OLIVIER IN DRACULA • ALSO STARRING DONALD PLEASENCE AND KATE NELLIGAN • A WALTER MIRISCH-JOHN BADHAM PRODUCTION • SCREENPLAY BY W.D. RICHTER • BASED ON THE STAGE PLAY BY HAMILTON DEANE AND JOHN L. BALDERSTON • FROM THE NOVEL BY BRAM STOKER • MUSIC BY JOHN WILLIAMS • SPECIAL VISUAL EFFECTS BY ALBERT WHITLOCK • EXECUTIVE PRODUCER MARVIN MIRISCH • PRODUCED BY WALTER MIRISCH • DIRECTED BY JOHN BADHAM • A UNIVERSAL PICTURE • PANAVISION · • © 1979 UNIVERSAL CITY STUDIOS. INC. ALL RIGHTS RESERVED

Index of stars

A selection of the leading stars to be found in the book. The index usually refers only to those films in which a star has been billed in the top three.

EMI

ONE MILLION YEARS BC

Starring
RAQUEL WELCH
JOHN RICHARDSON

Produced by
MICHAEL CARRERAS

Directed by
DON CHAFFEY